Connecting Literature and Science

This book presents a case for engagement between the sciences and the humanities. The author, a professional chemist, seeks to demonstrate that the connections between those fields of intellectual activity are far more significant than anything that separates them. The book combines a historical survey of the relationships between science and literature with a number of case studies that examine specific scientific episodes—several drawn from the author's own research—juxtaposed with a variety of literary works spanning a wide range of period and genre—Dante to detective fiction, *War and Peace* to *White Teeth*—to elicit their common themes. The work argues for an empirical, non-theory-based approach, one that is closely analogous to connectionist models of brain development and function, and that can appeal to general readers, as well as to literary scholars and practicing scientists, who are open to the idea that literature and science should not be compartmentalized.

Jay A. Labinger is the Administrator of the Beckman Institute at the California Institute of Technology. Trained as an organometallic chemist, he has published 200+ technical papers and patents and 20+ non-technical essays, along with books on the history of chemistry and sociology of science. He is a Fellow of the American Association for the Advancement of Science.

Routledge Interdisciplinary Perspectives on Literature

For more information about this series, please visit: www.routledge.com/
Routledge-Interdisciplinary-Perspectives-on-Literature/book-series/RIPL

Connecting Literature and Science

Jay A. Labinger

Routledge
Taylor & Francis Group

NEW YORK AND LONDON

First published 2022
by Routledge
605 Third Avenue, New York, NY 10158

and by Routledge
2 Park Square, Milton Park, Abingdon, Oxon OX14 4RN

Routledge is an imprint of the Taylor & Francis Group, an informa business

© 2022 Jay A. Labinger

Library of Congress Cataloging-in-Publication Data
A catalog record for this title has been requested

ISBN: 978-1-03-205350-9 (hbk)
ISBN: 978-1-03-212912-9 (pbk)
ISBN: 978-1-00-319718-8 (ebk)

DOI: 10.4324/9781003197188

Typeset in Sabon
by Newgen Publishing UK

to Andy and Barbara

Contents

Figures

Acknowledgments

I am greatly indebted to my fellow scholars in both the literature and science and science studies communities (they are far too numerous to list here; many are cited in the text) for inspiration, helpful suggestions, and/ or just general encouragement.

Special thanks go to two chemist colleagues, Roald Hoffmann and Steve Weininger. In addition to generously offering to read earlier versions of the manuscript and provide comments on both specific and global aspects, all of which helped point me towards a vastly improved final version, they have been invaluable guides and companions during my explorations into this fascinating field. I also thank Harry Collins for long-term guidance, as well as for reading and offering useful suggestions on Chapter 3; and Naomi Oreskes for permission to quote from an extended email correspondence on the same chapter.

I thank the following for permission to use previously published material:

- the first epigraph comes from *Pandora's Hope: Essays on the Reality of Science Studies* by Bruno Latour, Cambridge, MA: Harvard University Press, Copyright © 1999 by the President and Fellows of Harvard College.
- Chapter 5 is reprinted from Labinger, J. A., "Encoding an Infinite Message: Richard Powers's *The Gold Bug Variations*." Published in 1995 in *Configurations: A Journal of Literature, Science, and Technology* 3 (1): 79–93, with permission of Johns Hopkins University Press.
- Chapter 7 includes material from Labinger, J. A., "Metaphoric Usage of the Second Law: Entropy as Time's (Double-Headed) Arrow in Tom Stoppard's *Arcadia*." Published in 1996 in *The Chemical Intelligencer* 2 (4): 30–36, with permission of SNCSC.
- Figure 7.8 is adapted with permission from a) Bursten, Bruce E., and Roger H. Cayton, 1986. "Electronic Structure of Piano-Stool Dimers. 3. Relationships Between the Bonding and Reactivity of the Organically Bridged Iron Dimers $[CpFe(CO)]_2(\mu\text{-}CO)(\mu\text{-}L)$ (L = CO, CH_2, $C=CH_2$, CH^+)." *Journal of the American Chemical Society*

108 (26): 8241–8249. Copyright 1986 American Chemical Society. b) Bénard, Marc. 1979. "Molecular Orbital Analysis of the Metal-Metal Interaction in Some Carbonyl-Bridged Binuclear Complexes." *Inorganic Chemistry* 18 (10): 2782–2785. Copyright 1979 American Chemical Society. c) Jemmis, Eluvathingal D., Allan R. Pinhas, and Roald Hoffmann. 1980. "Cp$_2$M$_2$(CO)$_4$—Quadruply Bridging, Doubly Bridging, Semibridging, or Nonbridging?" *Journal of the American Chemical Society* 102 (8): 2576–2585. Copyright 1980 American Chemical Society.

- Figures 8.2, "Ascending and Descending," and 8.3, "Waterfall," are reproduced with permission of the M. C. Escher Company. © 2021 The M. C. Escher Company—The Netherlands. All rights reserved. www.mcescher.com.

- Figure 9.3 is reprinted with permission from Soai, Kenso, Takanori Shibata, and Itaru Sato. 2000. "Enantioselective Automultiplication of Chiral Molecules by Asymmetric Autocatalysis." *Accounts of Chemical Research* 33: 382–390. Copyright 2000 American Chemical Society.

Epigraphs

The only realistic way for a mind to speak truthfully about the world is to reconnect through as many relations and vessels as possible within the rich vascularization that makes science flow .... Pasteur too is exploring, negotiating, trying out what holds with what, who holds with whom, what holds with whom, who holds with what. There is no other way to gain reality.

Bruno Latour

Only connect! That was the whole of her sermon. Only connect the prose and the passion, and both will be exalted, and human love will be seen at its height. Live in fragments no longer.

E. M. Forster, *Howards End*

1 Introduction

About This Book

As a professional chemist, and an avid reader, I've been intensely interested in both literature and science for as long as I can remember. During much of my life I considered them pretty much as disjoint arenas—one for work and the other for play—somewhat akin to what Stephen Jay Gould called "non-overlapping magisteria." (He used the phrase to separate science and religion (Gould 1999); he emphatically did *not* apply it to science and the humanities (Gould 2003), as we shall see later on.) But I grew increasingly dissatisfied with that view: it seemed to me that the intellectual world doesn't need to work that way, and indeed does *not* work that way. For the last 25 years or so I have come to see how science and the humanities *do* in fact overlap greatly—to focus much of my attention on the "and" in literature *and* science (which I will henceforth abbreviate as L&S, although I will often take the "L" to refer to the humanities more broadly).

My initial interdisciplinary forays were of a fairly *ad hoc* nature: occasionally some literary work I happened to be reading—a novel, a play, an essay—seemed to evoke something familiar from my scientific work, or the converse. I can best describe this experience as a kind of *resonance*.[1] A typical phenomenon associated with that term is when ringing a bell causes another nearby bell that is tuned to the same note (or an overtone thereof) as the first to sound. In like manner, when an idea from one domain activates thought about something from a different domain, we call that resonance. By analogy, the perception of such a resonance would suggest that the two ideas have some basic underlying concept in common—that they are somehow *connected*. I became intrigued with that notion and set out to explore those connections and where they might lead, guided by no theoretical or methodological principles—at least, not consciously.

I soon became aware that this was far from an original pursuit; indeed, there was in fact an active L&S community, including a Society for Literature and Science. I began attending meetings regularly, and found there a congenial and stimulating group of interest-sharing scholars.

DOI: 10.4324/9781003197188-1

I was more than a little disappointed at how few of them came from the "S" side: a typical meeting would include no more than a handful of practicing scientists, although it is true that members on the "L" side often have solid scientific backgrounds, many having undergraduate degrees in a scientific field, for example. Nonetheless, I have greatly enjoyed and learned from those who gave presentations and participated in discussions at these meetings. I hope (and believe) that the converse has been true regarding my contributions.

Over the course of these years I gradually recognized a pattern or model (it would be too presumptuous to call it any sort of theory) in the approach I took in those contributions—talks given at meetings as well as occasional written pieces—that I thought might serve well as the basis of a discussion of L&S that would appeal to a broad potential audience, including:

- literary scholars who are active in (or even just aware of) L&S, and are interested in seeing how a scientist approaches the field.
- practicing scientists who are interested in literature (and/or the humanities more broadly) and in exploring relationships between their vocation and avocation, so to speak.
- general readers who have interests in the sciences along with their literary tastes, and don't necessarily believe those are or should be separate arenas.

This work is thus intended to contribute to that discussion, and to present a scientist's perspective on, and arguments for, engagement between the sciences and the humanities.

In Chapters 1–4 of the book I address general issues; the balance of the present chapter offers a mini-biographical sketch of how my L&S interests developed, and then considers the case for L&S as a worthwhile endeavor. Chapter 2 summarizes the history of L&S as a field of interdisciplinary scholarship, from the Huxley/Arnold debate of the late 19th century up to the present; Chapter 3 recounts the so-called Science Wars that erupted in the 1990s, which were characterized by manifestations of resistance to "science studies"—including L&S—that some took to be aimed at destabilization of scientific authority. Chapter 4 presents what I call a connectionist model for L&S, and attempts to both explain and justify it on several interlocking levels: the central role of metaphor in scientific as well as literary discourse; the function of analogy and how its exploration illuminates *both* sides of analogies under consideration; and the relation/connection/analogy between this model of L&S and connectionist models of brain development and function. It also briefly examines a couple of alternate approaches that have more of a scientific-theoretical underpinning.

I then present seven "case studies" in L&S, meant to illustrate and show the utility of this connectionist approach. The studies are largely independent—although there are several motifs that carry over from one to another in some cases: code links the first and second; translation, the second and third; creation of life, the fifth and sixth—and the inspirations that triggered them arose from a variety of sources. But each is centered upon a topic that evinces resonances between one or more literary works and scientific themes. Several of the latter are drawn from chemistry, not only because it is my own professional specialty, but also because it has received considerably less attention from the L&S and science studies communities than have physics and biological sciences.

The first study, Chapter 5, is a reprint of my initial effort in the field, a detailed study of a single book. I include it in unchanged form (I might well have written it differently today) as a sort of formative piece, an indication of how I was thinking when I first got involved in L&S, as recounted in the following section. Chapter 6 started with that same book: considering some of the aspects I hadn't covered, coupled with reading a more recent novel as well as a book on sociology of science, led me to new insights about one of my own scientific papers. Chapter 7, in contrast, was first inspired by a scientific paper, specifically a claim made therein with which I strongly disagreed. While trying to formulate a dissent on strictly scientific grounds, I recognized what seemed to me a parallel argument from the realm of literature. All the other studies had similarly varied geneses: Chapter 8 from attending a play; Chapter 9 again by a scientific paper; Chapters 10 and 11 by groups of works which—despite their considerable differences—gave me that feeling of resonance, in part because I happened to read them during a short period of time. (I recognize that serendipity has played a significant role in my L&S experiences!)

Lastly, in Chapter 12 I tie together the various themes of the case studies, show how they support my connectionist model, and advocate for continuing efforts along those lines.

A few general comments. First, I have made no attempt at being comprehensive, which would in any case be impossible. Many scholars from both the humanities and science camps have offered insightful commentary on these issues; I have had to be selective in my citations, and mean no disrespect to anyone who has been left out for reasons of space. In particular, while it is arguable that US scholars have often taken the lead in the L&S field (Rousseau 1978, 583), I suspect that I may have unjustly underrepresented scholars from other countries. I do offer some suggestions for "Further Reading" at the end of the book.

Second, I have a stylistic conflict. As an author, I like to interpolate digressive asides in my texts in the form of footnotes, but the publisher of the present work prefers the use of endnotes. As a reader, I strongly *dislike* repeatedly having to turn to the end of the book and back again just to see whether an endnote is sufficiently interesting and/or relevant to justify the time and trouble. (Reading *Infinite Jest* (Wallace 1996) a few years ago may be in part responsible for the intensity of that aversion.) Assuming that my potential readers may feel the same, I have chosen to incorporate most such commentary into the regular text, hopefully not too intrusively. A small number of endnotes have been added for some of the lengthier (and, perhaps, less essential) asides.

Third, the level of the scientific content presented in several chapters may be a little challenging to non-technical readers, but I believe most of it is important to my arguments, and that working through it will ultimately prove rewarding. In some instances (primarily in Chapters 6 and 9) I have given a simplified version of the chemistry, which should be sufficient for grasping the connections to the literary aspects of the subject, and have relegated the more complex details to an Appendix.

Lastly, I make rather liberal use of quotation. One of my aims is to highlight works—both of fiction and of scholarship—that strike me as particularly relevant and instructive; and a well-chosen quote, I believe, is often the best way to convey that. I have not gone so far as Walter Benjamin, who reportedly dreamt of producing a book consisting *only* of quotations (Arendt 1968)—his idea being, I presume, that if they were suitably selected and juxtaposed, no commentary at all would be needed—but I do follow the lead of another great essayist: "Je ne dis les autres, sinon pour d'autant plus me dire" (de Montaigne 1834, 69). My translation: If I speak in the voices of others, it is only to express myself better.

My Origins Story

Growing up in Los Angeles, I more or less decided on chemistry as my future career in my early teens—long before I had ever had any formal course in the subject. Why? I'm not entirely sure. I was certainly interested in science in general, and I know I received the usual (for boys, anyway, at that time) Gilbert chemistry set for one of my late-pre-teen birthdays. I remember enjoying working with it, but not particularly more so than some of the others of what we might now call STEM-related "toys"— a microscope, an Erector set, *etc.* Many of my colleagues in chemistry reminisce about their childhood passion for experimenting to see what could be done with chemicals—dramatic color changes, explosions—that inspired them to go into the field, but I do not recall any such intense fascination. What I *do* recall is reading, with great interest, several fairly elementary treatises on chemistry that I found in our house. They had belonged to my father, who had had thoughts of majoring in chemistry when he started college (University of Illinois) in the late 1930s, but was only able to complete a single year before he had to leave to help his parents financially. I don't know why he happened to keep those books around; to the best of my recollection he never explicitly encouraged me to go into chemistry.

On reflection it strikes me as interesting, and quite possibly significant, that my initial attraction to chemistry was more textual than experimental, a predilection that to a large extent has persisted. Even though for me chemistry is entirely an experimental science at its core—I was never tempted to pursue theoretical or computational work—I always enjoyed *thinking about* experimental results rather more than performing the physical work of obtaining them, and was more than happy to cede the latter to my students, postdocs and technicians when I reached the appropriate point in my career. (I haven't done an experiment with my own hands for 35 years or more.)

But texts did then and still do appeal to me strongly: I have always been an avid and eclectic reader. That included—as was practically *de rigueur* for a teenaged boy in those days—vast quantities of science fiction[2]: most of the works by major figures such as Asimov, Heinlein, and Clarke, along with a haphazard selection of books by other writers—whatever looked interesting on the shelves of the public library. I've forgotten most of them, but one that stuck in my mind—not for any reason having to do with quality—was *The Voyage of the Space Beagle* by A. E. van Vogt.

For those unfamiliar with that work (much of my readership, I suspect), it is a saga about a spaceship being outfitted for interplanetary exploration, with a crew carefully chosen to represent all fields of scientific expertise. They visit a number of planets (maybe four? I have deliberately chosen *not* to re-read the book, but rather to rely entirely on my possibly flawed memory), each of which is found to be inhabited by a different alien entity that presents a mortal threat to the survival of the expedition. In each case the collective talents of the team turn out to be inadequate to deal with the threat. Fortunately, though, one more member had been added at the last minute before departure—a representative of a new field called "Nexialism," a sort of interdisciplinary synthesis of all the other areas represented—and he is able to save the day each time. If I remember correctly, most or all of the traditional experts begin by ignoring, even resenting, the newcomer, but eventually (after the first couple of triumphs) they recognize his superior abilities and defer to him. What I definitely *do* recall is having been impressed by the idea that a generalist could outperform a team of specialists.

To be sure, I was a fairly naïve reader at that age (around 12), and many subtleties (and not-so-subtleties) sailed right past me. For example, it wasn't until a number of years later that I recognized how far Heinlein's political philosophy differed from my own; I simply read him as the author of apolitical space operas. In the case of *Space Beagle* (I didn't even appreciate the title then!), it was only after a similarly inordinate amount of time that I understood how I had been swindled. van Vogt's Nexialist triumphs over the specialists *not* because (or at least not only because) of any inherent advantage conferred by being a generalist, but rather because this *particular* generalist is made out to be a better chemist than the chemists, a better physicist than the physicists, a better biologist than the biologists, *etc.* (It was still later that I read somewhere that van Vogt had been a devotee of L. Ron Hubbard, and that "Nexialism" was intended to stand for a thinly-veiled version of Dianetics, the predecessor of Scientology.) Nonetheless, whether or not van Vogt deserves any of the credit (or blame) for it, respect for eclecticism/interdisciplinarity/generalism seems to have stuck with me through the years.

My attraction to chemistry survived high school mostly unperturbed. The curriculum, which included hefty doses of science, math, English, history, and a foreign language, was essentially dictated for any student contemplating attending the University of California. Even though I found many of the non-science courses more involving than some of those in science, including chemistry, I didn't change my mind; I just put that down to the quality of the teachers. When it came to choose a college, though, I picked Harvey Mudd College, which, more than most of the alternatives I had seen, placed great emphasis on balancing an intensely focused scientific education with a substantial amount (around one-third of the total requisite course load) of humanities and social sciences. Furthermore, as a member of the Claremont Colleges consortium, it provided much greater access to a wide range of faculty and subjects than an individual college of its size (only around 300 students *in toto* at the time) could possibly have permitted.

I did indeed receive an outstanding education at Harvey Mudd, both in chemistry (and science and math more generally) and in the liberal arts. I benefitted especially from two inspiring professors of literature, Benjamin Saltman and George Wickes. I recall getting back an essay from the former, marked up generally favorably, but ending with the comment that I wrote about literature "very scientifically." At the time I took that as a compliment, but on further reflection I'm not at all sure he meant it that way (I never asked him, alas). And in fact, the scientific and liberal arts curricula *were* kept pretty much separate, although at the time this didn't strike me as in any way surprising. The faculty did include one historian of science and technology, as well as a chemist who transformed himself into a philosopher of science; but there were few if any formal courses in those disciplines (I took none), and nothing at all that might have been called literature and science (which, to be sure, was hardly a common course designation back then).

So I left college, and went off to do graduate work in chemistry (at Harvard), much as I had come in: with strong interests in the humanities, especially literature, but still largely in split-brain mode: one would be my vocation, the other my avocation. And thus it remained, through five years as a PhD student, two years as a postdoc (Princeton), six years as a chemistry faculty member (Notre Dame), five years in industry (first Occidental Petroleum, then ARCO), and ultimately back to academia at Caltech, where I've been since 1986. Some of my recreational reading, to be sure, was located in the science/humanities borderlands. The book that affected me most strongly in that regard was Douglas Hofstadter's *Gödel, Escher, Bach* (henceforth *GEB*),[3] which appeared around 1980, although I didn't read it until some years later—daunted by its length and reputed (from reviews I *had* read) complexity. When I finally did, I was awed by his ability to draw *connections* between such nominally distinct realms—math, art, music (one of my particular passions)—and to be both entertaining and rigorous at the same time. The extensive wordplay (another passion of mine) was another attraction.

But *GEB* didn't inspire me to try my own hand at such cross-disciplinarity—in large part because I knew perfectly well that I could never hope to produce anything like it. I continued reading in recreational mode for the next few years, through the job changes and relocations outlined earlier, until around 1991, when I saw a review of a recent novel, *The Gold Bug Variations* by Richard Powers (an author of whom I had not been previously aware). The title alone, with strong flavors of both music and wordplay, was enough to hook me; and reading the book—which I completed in the course of a day or two, despite its 600+ page length—finally convinced me that just reading for fun was not sufficient. As with *GEB*, it was the intimate intertwining of disparate themes—in this case molecular biology, Bach's *Goldberg Variations*, coding in many guises—that impressed me, along with a heavy dose of puns (starting with the title), metaphors, and allusions—some but by no means all of which I recognized the first time through. I thought I understood much of what Powers was trying to do, as well as the often subtle devices he employed for his purposes, and felt—much as I often have felt on reading a complex paper on chemical mechanisms, for example—an urge to share my insights with others.

At this point it belatedly occurred to me that there *must* be others interested in the relations between literature and science, and a brief exploration on the Web (despite its relatively primitive state at the time) quickly turned up a rich listing of names and resources. Mostly because of geographical proximity, I picked out George Rousseau, who was then at UCLA, and readily identifiable as prominent in the field, for a first approach. He was most helpful and encouraging, and pointed me towards the (then-called) Society for Literature and Science—SLS—and in particular to another chemist, Steve Weininger, who had been one of the founders thereof. Pursuing those suggestions led to my participation in SLS (now SLSA, having added "the Arts" to its title) and friendship (and frequent collaboration) with Steve, both of which have continued at a high level throughout the ensuing quarter century. I attended my first SLS conference in 1993, presenting a paper on *Gold Bug* as part of a panel on Powers, and also meeting many other scholars who shared my interests—although, as mentioned earlier, only a handful consisted of practicing scientists, a situation that to my regret has persisted.

The SLSA journal *Configurations*, which was inaugurated just about the time I joined, has played a significant role in my evolution. Not only did it eventually publish my full-length paper on *Gold Bug* (Labinger 1995a; it is reproduced, in its original form, as Chapter 5 of this book), but also, the very first paper in the first issue—"What Are Cultural Studies of Scientific Knowledge?" (Rouse 1993)—introduced me to yet another arena of scholarship at the boundaries between science and the humanities, variously referred to as sociology of science, science studies, *etc.* I was particularly attracted by what *seemed* to present natural opportunities for joint interests, even collaborations, across the sciences/humanities divide; but on delving further into the field, I was disappointed by the quite limited extent to which that appeared to be taking place. I decided to express my feelings in the form of a somewhat polemical essay (Labinger 1995b), which I submitted to one of the leading journals in the field, *Social Studies of Science*. For once my timing was impeccable: the so-called Science Wars had recently become highly visible, and the journal's editors were looking for just such a piece as the nucleus for a more reasoned discussion of the issues involved. As a consequence I began interacting across disciplinary boundaries with yet another group, with interests not identical to but significantly overlapping with those of my SLS colleagues. Chapter 3 explores these matters more thoroughly.

I should also take note of a seemingly unrelated development, which coincidentally took place around the same time that I discovered *Gold Bug*. This was also a book-centered epiphany, but experienced by my wife Andrea rather than myself. She was in Mexico with a group of her students, and happened upon a book (*La Bobe*, by Sabina Berman) about a Jewish grandmother, which spoke to her so strongly that she felt compelled to translate it into English (Berman 1998), the beginning of her second career in literary translation. In following her work (and, occasionally, offering suggestions) I began to discern parallels between the sorts of problems that scientists and translators encounter, a perception that has contributed to my interests in, and approaches to, the pursuit of literature and science. A particular illustration of this aspect of my journey is the focus of Chapter 7.

Why Literature *and* Science?

The preceding section sets out why *I* have come to pursue L&S—and a personal interest in both areas is a *sine qua non* for participation; that is why I have made no attempt to suppress my personal voice in this book, hopefully without any serious detriment to reaching the audiences mentioned earlier. But clearly a case for L&S needs to be more universally convincing from a scholarly point of view: I offer those arguments here.

In the "Introduction" to the important *Cambridge Companion to Literature and Science*, Steven Meyer characterizes the field of L&S as "a dynamic platform for investigation into the many ways that the humanities and sciences share (1) a fundamentally pluralistic outlook; (2) common cultures and practices; and (3) a commitment to expanding the range and capabilities of empiricist approaches." He goes on to comment that "in a world increasingly mediated by technoscience it may be expected that many will wish to know more about how the sciences and humanities inform one another" (Meyer 2018c, 1). That stands in contrast to what is probably the default assumption for most of us (including me during my earlier career, as described earlier): that science and literature reside in distinct worlds. A common form of expressing that separation is something like "Literature is about words; science is about things," which is well represented in the chapter "Words and Things" in a book titled *Between Science and Literature* (Livingston 2006, 4–10). To be sure, humanists recognize that their words are often about things. Scientists will generally acknowledge that their things cannot be described without words, but that does not equate to admitting any sort of equal importance, let alone cohabitation. According to Nobel-winning chemist Roald Hoffmann, the common scientist's attitude (not his own, by any means!) is: "In science, we think that words are just an expedient for describing some inner truth, one that is perhaps ideally represented by a mathematical equation. Oh, the words matter, but they are not essential for science" (Hoffmann 2012, 39).

Here are two versions from literary scholars that express much the same point. First Kate Hayles, a leading L&S scholar, discussing a text by biologist and science popularizer Richard Dawkins:

Dawkins, a skillful rhetorician keenly aware of the value of a good story, nevertheless espouses what might be called the giftwrap model of language. This model sees language as a wrapper that one puts around an idea to present it to someone else. I wrap an idea in language, hand it to you, you unwrap it and take out the idea …. For example, at the critical juncture where the narrator is switching the unit of selection from the individual to the gene, we find this assertion. "At times, gene language gets a bit tedious, and for brevity and vividness we shall lapse into metaphor. But we shall always keep a skeptical eye on our metaphors, to make sure they can be translated back into gene language if necessary."

(Hayles 2001, 147)

and then Livingston, from his "Words and Things" chapter:

[B]oth scientists and humanists tend to overstate the independence of language from the world. Each begins by treating words and things as separate and then offers to connect them, though in rather different ways. Science, one might say, offers to nail words to things …. Ideally, language is conceived as a space of pure, undistorted reference to (or representation of) the world, rather like the controlled conditions of a scientific experiment …. One might even argue that an inability to see beyond the referential dimension of language is an asset for scientists, one that makes it easier to sustain belief in the scientific enterprise.

(Livingston 2006, 8)

I find it striking how different are the metaphors on which these two extracts are based. Hayles's scientist thinks words are readily detached from ideas, whereas Livingston's scientist wants them to be firmly nailed in place. But actually these nominally polar opposites point the same way. (It might be noted that Livingston acknowledges Hayles, who provided a foreword to his book, for support and inspiration.) Hayles and Livingston (like many other L&S scholars) here both reject Dawkins's claim that metaphoric language is mere window dressing that can be discarded at any time, leaving behind the completely unambiguous (in Dawkins's mind) "gene language."

Many scientists, as well, have been skeptical about the possibility of language-independent access to the "real" nature of things. A good example is found in chemist Ted Brown's response to the question of whether we can actually "see" atoms; note that he insists that even a "densely mathematical description" must be considered metaphorical:

> [W]e must remember that any model we might use to characterize the atom is metaphorical, whether it be that of a billiard ball, a plum pudding, a miniature solar system, a cloud of negative charge surrounding a positive center, or a densely mathematical description .... the images we obtain are indeed based on a stable, mind-independent reality .... One is moved to think, "Surely we are really seeing the atoms here!" What we see are constructs that at their best represent reliable models of reality, with sufficient verisimilitude to serve as productive metaphors. They facilitate correlations, predictions, and interpretations of other data and stimulate the creative design of new experiments. *That is all we can hope for.*
>
> (Brown 2003, 99; my italics)

Livingston cites Michel Foucault's *The Order of Things* as a direct antecedent to his own work (15–24), pointing out that Foucault's original title in French is *Les Mots et Les Choses*—literally Words and Things—changed by the publisher (with the author's concurrence) to avoid confusion with earlier books of that title (Foucault 1970, *viii*). In his "Foreword to the English edition" Foucault explains that one of his aims is to show that what he calls "semi-formal knowledge," dealing with matters such as living beings and languages, can and should be studied in the same manner as "rigorous sciences" such as mathematics and physics (*x*). He goes on to state that he wants "to reveal a *positive unconscious* of knowledge: a level that eludes the consciousness of the scientist and yet is part of scientific discourse, instead of disputing its validity and seeking to diminish its scientific nature" (*xi*). Here he anticipates, and attempts to deflect, an attack that was to become all too frequent (as we shall see in Chapter 3): that he and his followers are intent upon undermining scientific authority.

All of these commentaries (along with many others I could have chosen to quote) essentially dispute the notion that we can deal with the subject matter of science without concerning ourselves much about words, by means of purely mathematical representations and/or by constructing and using a language whose components correspond rigorously to empirical observables—*i.e.*, things. That was the goal of the logical positivists in the mid-20th century, and can be traced back even further to the concept of "carving nature at its joints." Such an idea, attributed originally to Plato (Slater and Borghini 2011), assumes that there are unambiguous rules for sorting and discussing things: the phrase "natural kinds" is frequently applied. It would follow that there must be correspondingly unambiguous rules for associating words with those things.

Foucault, of course, disagrees. In his Preface (*xv–xxiv*) he cites a *literary* source as his direct inspiration: an essay by Borges (1964a, 103), who quotes a "Chinese encyclopedia" (surely invented—the passage is so completely Borgesian!) in which

> animals are divided into: (a) belonging to the Emperor, (b) embalmed, (c) tame, (d) sucking pigs, (e) sirens, (f) fabulous, (g) stray dogs, (h) included in the present classification, (i) frenzied, (j) innumerable, (k) drawn with a very fine camelhair brush, (l) *et cetera*, (m) having just broken the water pitcher, (n) that from a long way off look like flies.

This fantastic taxonomy—which even violates rules of logic, as in the self-referentiality of item (h)—strongly connotes the impossibility of a unique and obvious classification scheme. While there certainly *are* things, there is no unequivocal way to classify them.

Steven Pinker provides another (though much less whimsical) illustration of the ambiguity inherent in classifying things and separating them from words—although I doubt he would see it as such:

> [I]t is a matter of fact, not opinion, that George W. Bush is the forty-third president of the United States, that O. J. Simpson was found not guilty of murder .... But though these are objective facts, they are not facts about the physical world, like the atomic number of cadmium or the classification of a whale as a mammal. They consist in a shared understanding in the minds of most members in a community, usually agreements to grant (or deny) power or status to certain other people.
>
> (Pinker 2002, 65)

I find this *very* strange in a number of ways, including the apparent implication that history (all of it?) is just a matter of what people agree upon. (There may well be commentators who *would* make such an argument, but I would not have expected Pinker to be among them!) Mainly, though, I am puzzled by the suggestion that the atomic number of cadmium and the classification of a whale as a mammal are facts ("things"?) about the physical world *in the same way*. Surely the latter cannot be completely separated from questions about how people have *agreed* to classify animals, even granting (which I certainly do) that "facts about the physical world" play an important role in how such agreement is reached.

In a similar vein, Livingston's book includes a rather capricious scheme (26–30) for arranging his library by "grammatical resemblance among their titles" to show how arbitrary such classification is. He concludes that "things are capable of being categorized in all kinds of ways ... this capacity seems to be as fundamental to the things of physics and chemistry as it is to language, culture and knowledge" (25). Such reasoning militates against the notion that words and things—or literature and science—are actually separable.[4]

Some scientists have been equally dubious about any essential disconnection between L&S. Notable among them is the late evolutionary biologist (and paleontologist, popular science writer, *etc.*) Stephen Jay Gould, whose last book (published posthumously) was subtitled "Mending the Gap Between Science and the Humanities" (Gould 2003). He makes several strong claims—all of which I fully endorse:

> An understanding of the social embeddedness of all aspects of science can forge an essential tie with humanistic studies and greatly aid the technical work of scientists as well.
>
> (116)

> A sympathetic application and understanding of "user friendly" themes in humanistic study will aid the approbation and acceptance of science by a suspicious general public. The breaking down of artificial barriers between the sciences and humanities will help even more.
>
> (138)

> The sciences and humanities have everything to gain (and nothing to lose) from a consilience that respects the rich, inevitable, and worthy differences, but that also seeks to define the broader properties shared by any creative intellectual activity, but so discouraged and so often forced into invisibility by our senseless (or at least highly contingent) parsing of academic disciplines.
>
> (258)

Other scientists who have expressed related opinions include Gould's former Harvard colleague—and frequent intellectual antagonist— Edward Wilson, and immunologist-turned-neuroscientist Gerald

Edelman; consideration of their contributions to the debate is deferred to Chapter 4.

Theoretical physicist (and popular science writer) Freeman Dyson explicitly mentions the "things *vs.* words" divide in the context of differentiating between science and theology (Dyson 1998), and I expect he would apply it to L&S as well. Nonetheless, he extols the synergistic value of combining scientific and humanistic study in a review of psychologist/behavioral economist Daniel Kahneman's 2011 book *Thinking, Fast and Slow*:

> There are huge differences between Freud and Kahneman .... Freud is literary while Kahneman is scientific. The great contribution of Kahneman was to make psychology an experimental science, with experimental results that could be repeated and verified. Freud, in my view, made psychology a branch of literature, with stories and myths that appeal to the heart rather than the mind .... The insights of Kahneman and Freud are complementary rather than contradictory. Anyone who strives for a *complete* understanding of human nature has *much to learn from both of them*.
>
> (Dyson 2011; my italics)

Granted, it might well appear easier to make a case for uniting literature with psychology than with, say, Dyson's own field. But it *has* been widely noted that recognizing ties between literature and physics—even if not rigorous—may provide valuable illumination. Here are just two examples, one from a physicist:

> But quantum physics is a theory of the microscopic, not of the human-sized .... while quantum effects in principle operate in living matter, they are negligible on the scale where biological processes occur. In short, quantum physics cannot be expected to give meaningful predictions or descriptions of human affairs. Its literary value lies in its metaphorical use and in the recognition that the essential ambiguity of the quantum has rich implications for the philosophical underpinnings of the physical world.
>
> (Perkowitz 2002, 365)

and one from a playwright:

> What the uncertainty of thoughts does have in common with the uncertainty of particles is that the difficulty is not just a practical one, but a systematic limitation which cannot even in theory be circumvented ... thoughts and intentions, even one's own—perhaps one's own most of all—remain shifting and elusive ... since ... the whole possibility of saying or thinking anything about the world, even the most apparently objective, abstract aspects of it studied by

the natural sciences, depends upon human observation, and is subject to the limitations which the human mind imposes, this uncertainty in our thinking is also fundamental to the nature of the world.

(Frayn 1998, 99)

These commentaries might appear to refer mostly—if not entirely—to one-way traffic between the domains: that while grasping scientific concepts may help inform humanistic matters, the reverse is improbable. But that, I believe, is *far* too simplistic. For example, consider quantum mechanics, which we understand profoundly in some senses and hardly at all in others, particularly in terms of any common sense meaning (Weinberg 2017). Might not bringing humanistic concepts to bear help deal with those puzzles? Paul Forman, a historian of science, has even suggested that the original development of quantum mechanics arose out of the same *Zeitgeist* that fostered 20th-century trends in literature and philosophy:

> [There is] overwhelming evidence that in the years after the end of the First World War but before the development of an acausal quantum mechanics, under the influence of "currents of thought," large numbers of German physicists, for reasons only incidentally related to developments in their own discipline, distanced themselves from, or explicitly repudiated, causality in physics.
>
> (Forman 1971, 3)

To be sure, claims of this sort have generated resistance, as we shall see in Chapter 3. I have found little unequivocal evidence that such ana-logical thinking has directly stimulated scientific progress, although there have been suggestions that understanding could potentially be influenced by extraphysical domains, such as music.[5] The possibility of *negative* influence—that inappropriate considerations could lead scientists down the wrong track—seems plausible as well. Such arguments have been applied to alchemists, in that their "willingness to use similarities between symbols and their referents ... in turn led to a belief in the causal powers of words and other symbols" (Gentner and Jeziorski 1993, 467). One could perhaps consider this amounting to delayed scientific progress (to the extent we want to call alchemy a science).

On the other hand, Werner Heisenberg, a central figure in 20th-century physics, concluded from his understanding of quantum mechanics that scientific and humanistic study are inextricably linked:

> [W]e can no longer view "in themselves" the building blocks of matter which were originally thought of as the last objective reality .... The familiar classification of the world into subject and object, inner and outer world, body and soul, somehow no longer quite applies, and indeed leads to difficulties. In science, also, the object

of research is no longer nature in itself but rather nature exposed to man's questioning, and to this extent man here also meets himself.

(Heisenberg 1958, 104–105)

Considerations of *narrative* offer yet another mode of expressing this linkage. We saw in an earlier quote (by Dawkins) the implication that narrative may be a part of, but is mostly superfluous to, science: it can easily be peeled away to reveal the "true" scientific facts. One commentator summarized that attitude (with which he does not at all agree) thus:

> Albert Einstein's contemporaries ... were not surprised to find a fair amount of storytelling in his popular accounts of relativity ... it is not that most of Einstein's readers saw narrative and scientific thinking and practice as fundamentally related. Rather, they saw storytelling as an auxiliary means to illustrate scientific ideas to the general public or sometimes scientists, while assuming science to be fundamentally different and, at least in principle, separable from narrative.
>
> (Plotnitsky 2005, 514).

Other authors have similarly focused on the role of narrative in science as primarily for communicating its findings, both to other scientists and the general public (Padian 2018; Olson 2015).

Today that view is not widely accepted: students of narrative much more commonly point to the *inseparability* of science from narrative. The introduction to a special issue of the journal *Studies in History and Philosophy of Science* suggests a number of ways in which narrative may be essential to scientific practice: by working to create coherence between apparently disparate aspects of a phenomenon or theory; by fostering understanding of complex systems by showing how they evolve over time; by drawing attention to contingencies, alternatives, counterfactuals, possibilities, causality, temporality, *etc*. All these are things that obviously feature in narratives; scientists could well benefit by recognizing their importance to their own thinking. As the authors put it: "Having a relevant theory does not substitute for having a rich narrative and having a good narrative may well embed a relevant theory: the two forms are not exclusive and may well reinforce each other" (Morgan and Wise 2017, 5).

Chemist-turned-author Primo Levi, whose widely ranging work eloquently demonstrates the power of narrative, offered a similar comment:

> I have frequently set foot on bridges that join (or ought to join) scientific culture with literary culture, crossing a crevasse that has always struck me as absurd. Some people wring their hands and describe it as an abyss, but then do nothing to bridge it; there are even those who work to widen it, as if the scientist and the literary man belonged to two different subspecies of humanity, speaking different languages, fated to ignore each other and incapable of cross-pollinating ....

I hope that these essays of mine, within their modest bounds of commitment and bulk, may show that there is no incompatibility between the "two cultures": instead, at times, when there is goodwill on both sides, there can be a mutual attraction.

(Levi [1985] 2015, 2014)

To sum up, fostering interactions between the sciences and the humanities appears attractive in a number of ways. Lowering interdisciplinary barriers could certainly improve the atmosphere in academia, which has become undesirably compartmentalized over the decades. For the general public, L&S may offer more familiar, less technically daunting approaches to the understanding needed for increased scientific awareness. Scientists may well find that exposure to the humanities helps not only to improve their ability to communicate with other scientists and non-scientists (Newman 2018; Clark 2019), but (at least) equally importantly, to get additional, fresh perspectives on their work. And that last, of course, works both ways: all of us, scientists and humanists alike, are devoting our energies to highly challenging problems, and we can use all the help we can get.

In that spirit I give the last word here to Aldous Huxley—novelist (*Brave New World* being the most science-related as well as best-known of his works) as well as grandson and grand-nephew, respectively, of T. H. Huxley and Matthew Arnold, two early L&S debaters from whom we shall hear in the next chapter. In 1963 he published a book-length essay addressing the past, present, and future of L&S. Although a subsequent commentator characterized Huxley's opinion of the field at the time of writing as exhibiting "little inspiration" (Woodcock 1978, 32), his concluding paragraph was quite the opposite: it can serve well as inspiration for the field in general, and my project in particular:

Thought is crude, matter unimaginably subtle. Words are few and can only be arranged in certain conventionally fixed ways; the counterpoint of unique events is infinitely wide and their succession indefinitely long. That the purified language of science, or even the richer purified language of literature should ever be adequate to the givenness of the world and of our experience is, in the very nature of things, impossible. Cheerfully accepting the fact, let us advance together, men of letters and men of science, further and further into the ever-expanding regions of the universe.

(Huxley 1963, 118)

Notes

1 The concept plays an important role in a number of aspects of science, usually involving some form of vibration or oscillation: resonance occurs when there is a matching of frequencies, as we will see illustrated later in this book.

2 It may be worth noting that SF does not play a very large role in this book, or indeed in my L&S interests in general. Granted, there is no really useful definition of what constitutes SF, as discussed at length by Margaret Atwood (2011), who has resisted having her work thus characterized; some of the works I consider later (including one of her own, *Oryx and Crake*) might well be associated with the term. But based on (for want of a better indicator) the sections of libraries and bookstores I tend to choose for browsing, my attraction to SF has waned over the years, roughly tracking—inversely—my entry into a scientific career. I did for a while continue devoting some of my recreational reading to SF, with tastes covering a wide range from "harder" (*e.g.*, Larry Niven) to "softer" (*e.g.*, Ursula LeGuin) authors, but by the time I started graduate school I kept up with the field—if it can be legitimately called such—only very occasionally. This is by no means to say that I do not consider SF to be an important part of L&S. Although Rousseau's early essay called SF an anomaly, in "a curious place in the recent evolution of the field" (Rousseau 1978, 588), in fact there has been a very substantial body of work devoted to that topic (see for example Milburn 2010; Dihal 2017; Stengers 2018; Markley 2018). Nonetheless, for whatever reason, my growing interest in L&S has not significantly nudged me back in that direction.

3 That all three of the eponymous figures appear in this book (Gödel in Chapter 11; Escher in Chapter 8; Bach in Chapter 5) is surely not coincidental, though I had no conscious awareness of connecting to *GEB* while working on those case studies.

4 The wealthy collector Abraham Warburg employed a less capricious scheme (although I suppose this entire section could be taken as an argument that capriciousness is in the eye of the beholder) for the organization of his private library in the early 20th century, with "works from a tremendous variety of disciplines and eras placed side by side in such a way as to suggest scarcely imaginable connections between them, potential similarities of approach, and lines of influence that seem inconceivable" (Eilenberger 2020, 130). In seeking such connections by juxtaposing (physically, in this case) ideas from different realms, Warburg is to some degree a precursor of my ideas here (although I became aware of him only after this book was essentially complete).

Eilenberger introduces Warburg's library as a major influence on the philosopher Ernst Cassirer (130–132), one of the four "magicians" (along with Martin Heidegger, Walter Benjamin, and Ludwig Wittgenstein) who comprise the protagonists of his study. Later he describes Cassirer's program in terms that call to mind several other passages I quote in this section:

> Cassirer saw the actual task of his philosophy as lying in the mutual illumination of as many "languages" as possible. Not just English, French, Sanskrit, and Chinese; as far as he was concerned, myth, religion, art, mathematics, even technology and law, were also languages.
>
> (293–294)

Generally I tend to avoid getting into matters of philosophy—like a (fictional) inquirer who appears elsewhere in this book, I "have not the philosophic mind" (Sayers 1936, 276)—but relating Cassirer to L&S might well constitute a productive area of exploration. Not now, though.

5 For example, Weinberg finds it possibly illuminating to draw an analogy between the somewhat mysterious concept of electron spin and a musical

chord (Weinberg 2017). Kepler attempted—entirely unsuccessfully—to relate the relative orbital speed of the planets to musical intervals (Bronowski 1965, 12). More recently, during the lead-up to the delineation of the periodic table, Newlands proposed an "octave rule" based on the observation that listing the lighter elements in numerical order by atomic weight results in a pattern of considerable resemblance in chemical and physical properties for every eighth entry (*e.g.*, lithium, entry 2, behaves very similarly to sodium, entry 9, and so on), and suggested that was somehow connected to the eight-note repeat pattern in the musical scale (C, D, E, F, G, A, B, C ...). It is not clear how seriously, if at all, he meant that to be taken; and in any case it was obviated, a few decades later, by the discovery of the noble gases, making the elemental resemblances recurring at every *ninth* entry (Scerri 2007, 76–79).

One *could* argue that music, or at least these harmonic aspects thereof, belong more properly to the realm of science than the arts; 19th-century physicist Hermann von Helmholtz made major contributions to understanding how we respond to music (Shapin 2019). But I don't find that entirely compelling. The "octave rule" in music surely does not follow in a completely deterministic, straightforward manner from the sciences of acoustics and the physiology of hearing, as the eight-note scale is central only to Western music, not universal to humanity. Armstrong offers a brief discussion of the question of physical/physiological *vs.* cultural factors in response to musical harmony, with some useful leading references (2013, 43–46).

2   A Brief History of Literature and Science

Huxley *vs.* Arnold; Snow *vs.* Leavis; Literature *vs.* Science?

While the phrases "literature and science" or "science and literature" can be found on the Google Books Ngram Viewer dating back at least to 1800, virtually all of the earliest entries are clearly meant to refer to two separate branches of knowledge or culture, rather than any conjunction between them.[1] L&S scholar Gillian Beer observed that interest in studying the relations between science and literature—paying real attention to the "and"—is a relatively recent development (Beer 1990, 783). The first such explicit consideration of the relationship between literature and science is commonly taken to be an exchange in the 1880s between biologist T. H. Huxley (who was widely known as "Darwin's Bulldog") and poet/literary scholar Matthew Arnold. But that episode, along with a reprise some 80 years later, reflects more an argument over supremacy than an attempt to effect any conjunction: the usage of "and" most resembles that in "The Hatfields *and* the McCoys."

In 1880 Huxley delivered an address titled "Science and Culture" (Huxley 1882) to mark the opening of a new institute, Josiah Mason's Science College, in Birmingham. In it he suggested—and lamented—that science had hitherto been generally left out of traditional education:

> How often have we not been told that the study of physical science is incompetent to confer culture; that it touches none of the higher problems of life; and, what is worse, that the continual devotion to scientific studies tends to generate a narrow and bigoted belief in the applicability of scientific methods to the search after truth of all kinds.
>
> (6–7)

The new college, in complete contrast, was to focus entirely on science, with no inclusion of "mere literary instruction and education," a proscription of which Huxley clearly approved:

DOI: 10.4324/9781003197188-2

> For I hold very strongly by two convictions—The first is, that neither
> the discipline nor the subject-matter of classical education is of such
> direct value to the student of physical science as to justify the expend-
> iture of valuable time upon either; and the second is, that for the pur-
> pose of attaining real culture, an exclusively scientific education is at
> least as effectual as an exclusively literary education.
>
> (7)

By this time the Industrial Revolution had considerably transformed
British society, but it had not had much impact on accepted standards
of what constituted high culture and learning. In that context, Huxley's
oration seems clearly intended as a declaration of war, being saturated
with martial language:

> the long series of battles, which have been fought over education in a
> campaign which began long before Priestley's time ... complicated by
> the appearance of a third army, ranged round the banner of Physical
> Science ... [which] must be admitted to be somewhat of a guerilla force.
>
> (2–3)

and also taking a jab at Arnold, "our chief apostle of culture," for writing
things that give support to the anti-science "Philistines" even while
professing "true sympathy with scientific thought" (8).

Arnold did not take long to reply. In 1882 he was invited to Cambridge
as the Rede Lecturer, a lecture series that dates back to the 17th century.
The large majority of the annual talks from 1859 to 1899 were on scientific topics (Huxley gave one in 1883, on evolution), but Arnold's talk,
subsequently printed in the journal *Nineteenth Century* (Arnold 1882),
was on "Literature and Science." He began by describing the relative
status of the fields, in terms quite antipodal to Huxley's:

> Ten years ago I remarked on the gloomy prospect for letters in this
> country, inasmuch as while the aristocratic class ... was totally indif-
> ferent to letters, the friends of physical science on the other hand, a
> growing and popular body, were in active revolt against them ....
> If the friends of physical science were in the morning sunshine of
> popular favour even then, they stand now in its meridian radiance.
> Sir Josiah Mason founds a college at Birmingham to exclude "mere
> literary instruction and education"; and at its opening a brilliant and
> charming debater, Professor Huxley, is brought down to pronounce
> their funeral oration.
>
> (216)

He even anticipated the concept of dividing the world into matters of
things and matters of words—the legitimacy of which was to become a
significant topic for debate, as we saw in the previous chapter:

This reality of natural knowledge it is, which makes the friends of physical science contrast it, as a knowledge of things, with the humanist's knowledge, which is, say they, a knowledge of words.

(222)

Overall, one would have to say, Arnold's tone and stance sound considerably more conciliatory than Huxley's. While making it clear which side he would support if it came to bellicosity:

If then there is to be separation and option between humane letters on the one hand, and the natural sciences on the other, the great majority of mankind, all who have not exceptional and overpowering aptitudes for the study of nature, would do well, I cannot but think, to choose to be educated in humane letters rather than in the natural sciences. Letters will call out their being at more points, will make them live more.

(228–229)

he was confident that any *apparent* conflict could be averted by accepting a broader understanding of "literature":

All knowledge that reaches us through books is literature ... when I proposed knowing the best that has been thought and said in the world ... I certainly include what in modern times has been thought and said by the great observers and knowers of nature. There is, therefore, really no question between Professor Huxley and me as to whether knowing the results of the scientific study of nature is not required as a part of our culture, as well as knowing the products of literature and art.

(220–221)

Indeed, many 20th- and 21st-century commentators and critics have taken the path of engagement. But before turning to that, let's look at another familiar debate—an example of history repeating itself. Again, a Rede lecture played a major role—in this case, the opening shot. It was delivered in 1959 by C. P. Snow, with the title "The Two Cultures and the Scientific Revolution." Snow had been trained in science, in which he had a very short-lived (and not very successful) career, but became much better known as a novelist and civil servant. He suggested that little had changed in the 80 years since Huxley and Arnold, although he *seemed* more ready to place blame on both sides:

Literary intellectuals at one pole—at the other scientists, and as the most representative, the physical scientists. Between the two a gulf of mutual incomprehension—sometimes (particularly among the young) hostility and dislike, but most of all lack of understanding.

(Snow [1959] 1998, 4)

along with a particular feature of the British educational system:

> I said earlier that this cultural divide is not just an English phenom-
> enon: it exists all over the western world. But it probably seems at its
> sharpest in England, for two reasons. One is our fanatical belief in
> educational specialisation, which is much more deeply ingrained in
> us than in any country in the world, west or east.
>
> (16–17)

Many have suggested that Snow's commentary was more reasonable—
less of a claim for scientific superiority—than Huxley's. Stephen Jay
Gould, for example, refers to "his utterly inoffensive and, in retrospect,
rather dull Rede Lecture" (Gould 2003, 89). This perception probably
has several origins, including the fact that Snow entered the debate
with credentials on both sides. Also, for a good part of the lecture
he turned to a less contentious issue: how science could help ameli-
orate poverty in the world. Looking back upon the episode four years
later, he regretted not having made that his title and main focus (Snow
1963, 79).

On closer examination, though, it is not so clear that Snow deserves
much credit as a conciliator. Even in the better-known 1959 essay there
are comments that come across at least as polemic as those of his forebear
Huxley:

> It isn't that [scientists are] not interested in the psychological or
> moral or social life. In the social life, they certainly are, more than
> most of us. In the moral, they are by and large the soundest group
> of intellectuals we have (13). Intellectuals, in particular literary
> intellectuals, are natural Luddites.
>
> (22)

and some even more so, as when he quotes (approvingly) an anonymous
scientist who associated "most of the famous twentieth-century writers"
with "bring[ing] Auschwitz that much nearer" (7). A recent book (Marx
2018) places Snow squarely within a millennia-old tradition of "hatred
for literature," pointing out that the Rede lecture grew out of an earlier
and more aggressive essay, also titled "The Two Cultures" (Snow 1956),
that *explicitly* proclaimed the moral superiority of scientists—even indul-
ging in language that would now surely be considered unacceptably
homophobic:

> It is that kind of moral health of the scientists, which, in the last few
> years, the rest of us have needed most; and of which, because the two
> cultures scarcely touch, we have been most deprived.
>
> (414)

[The tone of scientific culture] is, for example, steadily heterosexual .... About the whole scientific culture, there is an absence—surprising to outsiders—of the feline and oblique.

(413)

But the main reason why Snow's strong comments—which (*pace* Gould) do not sound to me so "utterly inoffensive"—have largely been overlooked and/or excused was a highly intemperate response issued by F. R. Leavis, then Cambridge University Reader in English. At his 1962 Richmond Lecture given at Downing College, and subsequently reprinted in *The Spectator* (Leavis 1962), Leavis tore into Snow on every ground imaginable, professional and personal:

> If confidence in oneself as a master-mind, qualified by capacity, insight and knowledge to pronounce authoritatively on the frightening problems of our civilization is genius, then there can be no doubt about Sir Charles Snow's .... But of history, of the nature of civilization and the history of its recent developments, of the human history of the Industrial Revolution, of the human significances entailed in that revolution, of literature, of the nature of that kind of collaborative human creativity of which literature is the type, it is hardly an exaggeration to say that Snow exposes complacently a complete ignorance. The judgment I have to come out with is that not only is he not a genius; he is intellectually as undistinguished as it is possible to be.
>
> (90)

He went on to sneer at both Snow's literary and scientific credentials: "as a novelist he doesn't exist; he doesn't begin to exist" (92); "The only presence science has is as a matter of external reference, entailed in a show of knowledgeableness. Of qualities that one might set to the credit of a scientific training there are none" (93).

Finally, though, Leavis moved past *ad hominem* attacks—echoing Arnold at 80 years' remove—to make a case for the importance of literature without diminishing the importance of science:

> But there is a prior human achievement of collaborative creation ... one without which the triumphant erection of the scientific edifice would not have been possible: that is, the creation of the human world, including language .... It is in the study of literature ... that one comes to recognize the nature and priority of ... the realm of that which is neither merely private and personal nor public ... it is something in which minds can meet.
>
> (100)

If we can bracket out the "my field is more important than yours" along with the personal insults, we see indications in both exchanges of the possibility of conciliation and cooperation from all four participants. There is even a suggestion of how to get there in Snow's (1963) revisitation, in which he drew attention to the emergence of what he considered a potential "third culture"—social studies—that might be capable of effecting a reunification of the first two (69–71). Bridging social studies and science—a field generally described as "science studies"—did indeed grow to become a major focus of scholarship in the next few decades. Unfortunately, as we shall see in the next chapter, much of that effort seems to have worked in the wrong direction, generating suspicion and hostility rather than reconciliation. Before turning to that story, though, let us examine the origins and evolution of literature and science as a distinct topic of scholarship.

The Emergence of L&S

In his introduction to the recent *Cambridge Companion to Literature and Science*, editor Steven Meyer proposes to divide the history of L&S as a field of scholarship into two "waves," running respectively from around 1945–1980 and from 1980 to the present (Meyer 2018c, 3), while also noting important earlier contributions, particularly Whitehead's *Science and the Modern World*. Some of Whitehead's comments seem a bit jarring, such as "[In] ordinary literature, we find, as we might expect, that the scientific outlook is in general simply ignored. So far as the mass of literature is concerned, science might never have been heard of" (Whitehead 1925, 111), which is embedded in an entire chapter (109–138) analyzing and comparing how Romantic poets such as Wordsworth, Shelley, and others reacted to science! Unquestionably, though, Whitehead's work has figured greatly in contemporary L&S (Meyer 2018b), although that influence took a while to take hold.

A large share of the credit for getting serious L&S scholarship underway has been attributed to Marjorie Hope Nicolson (Beer 1990, 789; Rousseau 1978, 583). Nicolson was Professor of English, first at Smith College and then at Columbia University, where from the mid-1930s on she offered a course on "Science and Imagination" (Nicolson 1965, 177), along with writing extensively in the field. Her work, and that of most of her contemporaries, falls squarely into Meyer's "first wave," which has been characterized as "the field ... in which scientific references in literature are documented" (Rousseau 1978, 584); explaining literature "by exposition of the scientific theories or discoveries to which it alludes" (Beer 1990, 789); or "trac[ing] the history of scientific concepts within literary contexts" (Meyer 2018c, 4). Basically, in other words, it consists of the exploration of science *in* literature, primarily in a historical context.

The *institutionalization* of L&S as a scholarly field can be dated, fairly precisely, to the late 1930s. During the business part of the 1937 meeting of the Modern Language Association (MLA) section on Philosophy and Literature of the Classical Period—not coincidentally chaired by Nicolson—it was agreed that "The interrelations of science and literature in the early eighteenth century" should be the topic for the 1938 meeting (MLA 1937, 1356). Such a session duly took place, with three talks: an Introductory "Survey of present tendencies in the field" given by chair Nicolson, followed by "Edward Tyson: Anatomist, Physician, and Litterateur of the 17th Century" and "The Educational Theories of Joseph Priestley" (MLA 1938, 1344–1345). An MLA Division for Literature and Science was officially inaugurated in 1939 (Meyer 2018c, 2), with a session titled "Relations of Literature and Science"; it offered four papers (two of which, interestingly, looked back to Huxley and Arnold): "Wordsworth and the Interpretation of Nature"; "Huxley and the Victorian Worship of Science"; "Matthew Arnold and Science"; and "Donne's Knowledge and Use of Paracelsus." Those present at the inaugural session (reportedly numbering 203: a rather impressive attendance!) further "approved unanimously the resolution of the Coordinating Committee that the literary curricula of colleges should include adequate representation of the literature based upon and connected with science" (MLA 1939, 1357).

A considerably stronger exhortation was offered several years later by anthropologist M. F. Ashley-Montagu:

> It should be obligatory for every department of English or Literature, and every department of science, to have at least one full course in the history of science and culture which every student for a degree must, and which every member of the department should, take.
>
> (MLA 1943, 24)

That year's session also included what sounds like a more forward-looking talk: the abstract of a paper titled "Literature and Science as Communication" ended with

> It is important that students of literature who wish to "defend the humanities" in an intellectual world dominated by scientific assumptions (1) analyze the strategic direction of current discussions of linguistic communications, and (2) organize analyses which will be suitable for the study of scientific communication and the study of literature.
>
> (MLA 1943, 23)

Still, the majority of papers given in the one or two L&S sessions held at subsequent annual MLA meetings rarely ventured outside of the "science in literature" paradigm. There was not much obvious broadening of the scope of the field—either quantitatively or qualitatively—until around 1980, where Meyer locates the beginnings of a "second wave" of L&S.

Indeed, the topics at an L&S session of the 1980 MLA meeting—"The Tight Fist of Philosophy, the Open Hand of Science, and Some Guesses about the Literary Methods of Chaucer and Langland," "The Use and Misuse of Giambattista Vico: Humanist Rhetoric, Orality, and Theories of Discourse," "The Orientalization of Western Discourse: Some Implications of Michel Foucault's Theory of Language," "Historical Periods and Cultural Change: Preconceptions and Preconditions," and "Toward a Period Theory of Discourse?" (MLA 1980, 995)—look *quite* different from those of the 1938 talks listed earlier. (It should be noted, though, that the 1980 meeting also featured a session on "Einstein and Modern Literature," more along the lines of the earlier "science in literature" approach.)

Meyer (2018c) perceives two main drivers of this development, both of which began to emerge around the late 1960s. First, the field of literary scholarship became increasingly theory-laden, strongly influenced by the likes of Derrida, Foucault, *etc.*, a tendency that has certainly been manifested within L&S, although perhaps one that has been waning more recently (see for example Lodge 2004). I will not have much to say about literary theory in general, but I will (in Chapter 4) look at approaches to the study of literature that are substantially based on concepts *from* scientific fields, specifically evolution and cognitive science.

The other was the appearance and rapid growth of a new (not entirely, of course) field that has generally been designated as "science studies." Its practitioners, who come from a wide variety of disciplinary backgrounds, take the position that the operations and findings of science can be examined from a critical perspective—even by those who are not professional scientists—rather than having to be taken as given fact. That stance (which occasioned a good deal of push-back from some scientists and more traditional philosophers of science, as will be discussed at greater length in the following chapter) characterizes a good deal of post-1980 L&S,[2] as described by (historian of science *and* L&S scholar) James Bono:

> From the early 1980s on, a new generation of scholars ... in the reconfigured field of literature and science [turned their] attention to the ... literary and linguistic dimensions of scientific practice and communication ... and ... the networks of exchange that foster the circulation of the objects, material practices, and epistemic things that contribute to the making of scientific knowledge.
>
> (Bono 2010, 556–557)

This is a picture of a much more symmetrical L&S—literature in science as much as science in literature—that many have argued for (Dillon 2018). In the examples from the 1980 meeting listed earlier we can see "attention to the literary and linguistic dimensions" in aspects such as the role of metaphoric language in scientific explanation, rhetorical examination of scientific discourse, *etc.* One of the pioneers of this form of study was Gillian Beer, whose *Darwin's Plots* (1983) initiated a large body of "work back and forth across the divide between science and literature, showing *both* how Darwin's science impacted literary form *and* how that science was, in turn, shaped by the literary examples Darwin had to draw from" (Griffiths 2018, 66–67). Many questions in L&S have been addressed in similar fashion. For just one more example, from my own field, the "chemical revolution" brought about largely by Lavoisier in the late 18th century has been credited as much to his introduction of novel nomenclature and modes of discourse as to the experimental facts he established (Golinski 1998, 117–119). A useful survey of critical studies in L&S, organized primarily by topic, may be found in Willis (2015).

From an organizational viewpoint, the next important development was the establishment of the Society for Literature and Science (SLS; the name was changed some 20 years later to add "the Arts," hence now SLSA). This society was launched at a History of Science Congress in 1985 (Weininger 1989, *xiii*)—not at an MLA meeting. But interest within MLA had unquestionably been equally strong at the time: the program for the 1985 meeting lists no fewer than *nine* sessions under the theme "Science, Technology and the Humanities," only three of which were formally sponsored by the Division of Literature and Science (MLA 1985, 900). There continues to be a strong component of historians of science in the field, most notably recognized in a "Focus section" in *Isis*, the journal of the History of Science Society, which was devoted to L&S-related articles (Bono 2010).

Chemist Steve Weininger, one of SLS's founders and its second president, later summarized the main goals of the new society as expansion of the scope of L&S studies—reflecting the turn towards a second wave which was already well underway within MLA—along with more substantial participation by practicing scientists (which was decidedly *not* happening at MLA):

> Literature and science has existed as a field of study since the 1920s (*sic*), when the Modern Language Association established a division of that name. Its practitioners were almost solely literary scholars, and its reigning paradigm was the "influence" model that focused on the one-way interaction from science to literature. By the 1980s there was a strong desire to open the field to a greater number of disciplines and approaches. Discussions among a small group of scholars envisioned a new Society for Literature and Science (SLS) where scholars from a broad range of fields, and particularly the sciences, would feel welcome, and where the discursive arena would belong to no single discipline or group of disciplines.
>
> (Weininger 2001)

The first goal was quickly met: SLS (later SLSA) has held annual meetings in North America since 1987, and started its own journal, *Configurations*, which published its first issues in 1993. A broad range of areas of interest has been represented in both the meetings and the publication from the outset. In response to strong overseas interest and participation, an additional series of European meetings was instituted in 2000, which eventually resulted in creation of a European Society (SLSAeu); and a British Society for Literature and Science was established in 2005. Both of these maintain close ties to SLSA while sponsoring their own annual meetings; there is also a Britain-based (online) journal, *Literature and Science*. It must be acknowledged, though, that attracting scientists has proven more challenging: active scientists have never accounted for more than a handful of members, publication contributors, or meeting attendees (Labinger 2017). Movement in that direction—however small—is one of the aspirations of *this* book.

L&S Today

In 1978 George Rousseau, a scholar of considerable stature in the field of L&S, offered a rather pessimistic assessment of its future. In particular, he feared that what he called the "intrusion" of literary theory—particularly as represented by structuralism and Foucault—was likely to be a death blow:

> Furthermore, an impression was given that structuralists were finally turning literary criticism into a science. This claim was not new, but many literary critics, believing themselves the champions of an ancient humanism now threatened by a science (more accurately, a pseudo-science, as they came to view structuralism) resisted the temptation to make a science out of literary criticism. It now appeared to some that structuralism in the 1970s would finally end the debate about the validity of literature and science as a field. Others, discouraged, abandoned the ship before it could sink. As proof, the membership of the Literature and Science Division of the MLA took a sharp plunge from which it has still not recovered. By 1975 there were so few members of the division that the Executive Council dissolved it; it now exists on a probationary basis, but its future is very much in doubt.
>
> (Rousseau 1978, 589)

That seems diametrically opposed to the positive role Meyer ascribed to theory in the evolution of L&S, and indeed Rousseau's doubt was *not* borne out in fact: the Literature and Science Division of the MLA recovered, to the extent that it recently (2013) claimed 2889 members (MLA 2015)! SLSA likewise has unquestionably been a successful venture, with steady growth from its inception: a recent (2018) meeting in Toronto featured around 135 sessions and 530 presenters, from an *extremely* broad range of fields (Society for Literature, Science and the

Arts 2018), almost triple the size of the first meeting I attended in Boston in 1993.

The calls for L&S coursework at early MLA meetings have been realized to a certain degree; as several commentators have pointed out, what we might call the "STEMification" of contemporary American higher education may well impart greater urgency to such calls (Droge 2017; Nash 2017). To the best of my knowledge there are no formal L&S departments or degree programs in any college or university in the US, but many have regular or at least occasional course offerings in the field, and some have organized faculty interest groups.[3] In Europe, in contrast, there *are* several formal programs, notably the Center for Literature and Natural Science at the University of Erlangen, an "inter-disciplinary research-centre ... dedicated to the reciprocal transfer of knowledge between physics and literature ... concerned with the import-ance of language and metaphors in physical research as well as with dis-cursive and narrative modulations of scientific theories in literary texts." The Center, which comprises faculty from both literature and science departments, sponsors courses, seminars, and workshops as well as a book series (ELINAS, n.d.). Several other L&S-focused research groups include a "Literature and Science Hub" (University of Liverpool, n.d.), a "ScienceHumanities" initiative (Cardiff University, n.d.), and a "Fiction Meets Science" program (University of Bremen, n.d.).

In retrospect, the expansion of L&S appears to have been virtually inevitable, given the increasing prominent role of science and technology in post-war society, especially (in the US) following the 1957 wake-up call of Sputnik. In such an environment, a scholar who declined to consider science might well feel the risk of starting to seem irrelevant. Furthermore, it would be hard to sustain the mere tracing of scientific themes and concepts within literature as an exciting pursuit for very long.

That trend applies equally to the creative side of L&S. In the pre-vious chapter we saw Aldous Huxley's enthusiasm about the *potential* of science in literature. While he was not so much impressed by what he saw as the accomplishments to date (writing in 1963, just a few years after Sputnik):

> [on poetry]: That the poetry of this most scientific of centuries should be, on the whole, less concerned with science than was the poetry of times in which science was relatively unimportant is a paradox ... the very fact that this is an age of science has relieved poetry of the need to have much direct and detailed scientific reference.
>
> (61)

> [on theater]: Science is a matter of disinterested observation, unpreju-diced insight and experimentation, patient ratiocination .... Passion and prejudice are always able to mobilize their forces more rap-idly and press the attack with greater fury; but in the long run ...

enlightened self-interest may rouse itself, launch a counterattack and win the day for reason. In the fictional world of the drama, this is not likely to happen.

(68–69)

[on literature in general]: [I]t is worth remarking that men of letters are ready to work very hard on obscure subjects of a nonscientific kind, but are not prepared to invest a comparable amount of labor in the artistic transfiguration of intrinsically less obscure scientific raw materials .... (101) The hypotheses of modern science treat of a reality far subtler and more complex than the merely abstract, verbal world of theological and metaphysical notions ... this reality is non-human, essentially undramatic, completely lacking in the obvious attributes of the picturesque. For these reasons it will be difficult to incorporate the hypotheses of science into harmonious, moving and persuasive works of art.

(106–107)

it should be obvious by now that Huxley's pessimism, like Rousseau's, was unwarranted. In the last half-century, novelists, dramatists, and poets have embraced scientific themes, examples of which will be encountered in the following case studies. *Many* more have been the subjects of L&S literature; a useful online listing has been provided by the "Fiction Meets Science" program (University of Bremen, n.d.). This interest extends even to the more mundane aspects of the scientific life. Gillian Beer commented—or predicted?—that

We shall not find in literature widespread reference to the ordinary doings of the sciences. We shall look in vain for novels and plays set in the laboratory, unless it is a laboratory at the moment of war .... So when we look for the 'scientist in literature' we shall not find him or her so much at the level of social description as at that of myth.

(Beer 1990, 794–795)

Au contraire: there is a substantial body of such work, sufficient to engender its own extensive website "dedicated to real laboratory culture and to the portrayal and perceptions of that culture—science, scientists and labs—in fiction, the media and across popular culture" (LabLit. com, n.d.).

How did Huxley get it so wrong? He completely failed to recognize what science studies have made clear (although many of us did not need science studies to point this out!): the "reality" of science is *anything but* "nonhuman" and "undramatic." Just take one well-known example from the realm of non-fiction (or at least partly so): James Watson's 1968 book about his and Crick's elucidation of the structure of DNA. What could be more human, more dramatic, than the interpersonal rivalries in his tale of the race to the answer? Any creative literary person familiar with stories such as these—produced by both active participants and subsequent commentators—should never be at a loss to spin "harmonious, moving and persuasive works of art" out of scientific threads.

Notes

1 This search was restricted to English, a possibly distorting preference that was acknowledged in the preceding chapter. On the other hand, the separation of literature and science may well have been more prevalent, historically, in English thinking. One commentator notes that, according to the OED, having a word "science" to refer to what we now call the natural sciences, as opposed to knowledge more broadly, is a usage primarily of English origin (Collini 1998, xi).

2 A review of the *Companion* challenged Meyer's account—and indeed a substantial part of the entire volume—for focusing far too much on the overlap between L&S and science studies (Willis 2018). Certainly, "second wave" L&S by no means *displaced* "first wave" activity, which has continued as a vibrant component of the field; but I tend to agree with Meyer about the important role of science studies in shaping its evolution.

3 Web searches on "literature and science programs" and permutations thereof turned up only a few hits that fit the description. Of course, many universities are centered around a College of Letters and Science, or Arts and Science, or some other similar title (at the University of Michigan it is LSA, for Literature, Science and the Arts—coincidentally identical to SLSA), but these organizational designations do not carry any particular emphasis on interdisciplinary overlap. Several of the hits led to university website pages headed "Literature and Science" that described interest groups, such as the University of Notre Dame (University of Notre Dame, n.d.), or just courses with that title. A search of course listings at a dozen or so colleges and universities (chosen mostly at random) found that each offered at least one course titled Literature and Science or something more specific (*e.g.*, Literature and Medicine, Literature and the Environment, *etc.*), with nearly all listed under the Department of English. One might think that courses taught *jointly* by faculty from humanities and science departments would be an attractive approach. I have heard a small number of such described at SLSA meetings, but they appear to be fairly rare.

3 The Science Wars

The Rise of Science Studies

I ended the previous chapter by focusing on the human aspects of the scientific enterprise. Very few, if any, commentators would deny that those have any place at all, but towards the end of the last century, particularly during the 1990s, the perception that they were being granted *too much* importance provoked a strong backlash—the Science Wars—in which L&S became enmeshed as part of the science studies that had played a key role in its evolution. In this chapter I briefly trace the history of this episode and some of the major issues it entailed, and attempt to show why it does *not* constitute any reasonable argument against the validity and value of L&S—even though some of the bellicosity persists to the present day.

The study of science from perspectives of the humanities is of course not a new development. History of science has been a recognized professional discipline in its own right since the early 20th century; a key figure was George Sarton, who was primarily responsible for both the journal *Isis*, first published in 1912, and the History of Science Society, founded in 1924 (History of Science Society n.d.). Philosophy of science in some sense can be traced back to Aristotle, but it also became institutionalized as a distinct subfield of philosophy in the 20th century: the Philosophy of Science Association was established in 1933 (Philosophy of Science Association n.d.). Credit for the emergence of sociology of science is generally awarded to a single scholar, Robert K. Merton, whose combined academic backgrounds in sociology (as an undergraduate) and history of science (as a graduate student with Sarton) were instrumental in his establishment of this new field in the 1930s (Holton 2004). But the extent to which any of these movements really constituted a critique was limited. A (considerably too simplistic) characterization of early workers in the three fields might be that they were happy to take scientific accomplishments as natural, not requiring explanatory analysis, and to content themselves with recounting the chronology, accounting for the logic, and describing the organization and institutions of the scientific enterprise, respectively.

DOI: 10.4324/9781003197188-3

The main starting point for the turn to contemporary practices in these fields (a useful summary may be found in Golinski 1998) is generally identified as Thomas Kuhn's (1962) book *The Structure of Scientific Revolutions*, although earlier workers clearly had begun to point in those directions. The most notable of these were Ludwik Fleck, who in the 1930s argued for the importance of cultural influence—"thought style" or "thought collectives" (Shapin 1980); and Michael Polanyi, who wrote in the 1940s about the role of personal commitments and "tacit knowledge" in scientific practice (Wigner and Hodgkin 1977, 430–433). Neither of these important contributors was very widely known before Kuhn, who acknowledged his debt to both of them (Kuhn [1962] 1970, *vi*, 44).

Kuhn began his career as a hard-core physicist, earning a PhD at Harvard, but soon shifted his attention to issues in history and philosophy of science. In *Structure*, Kuhn argued that the historic evolution of science could not reasonably be described in terms of linear progress; rather, there have been periodic dramatic changes ("revolutions") in the dominant mode of thinking about any given scientific field. He popularized (at least) two terms: "paradigm," to describe that dominant mode; and "incommensurability," the idea that the central concepts in one paradigm cannot be carried over into its successor, making impossible a completely objective assessment of the relative merits of the two—it depends upon which paradigm one is committed to.

These ideas have been suggested to imply two dreaded *isms*: social constructivism, the position that scientific facts are not simply out there in the world, waiting to be gathered up by observation and experimentation, but are, rather, constructed by scientists carrying out their work; and relativism, the claim that knowledge is never absolutely true, but only so within its particular context. Kuhn himself denied that his position entailed relativism (Kuhn [1962] 1970, 205–207), but it is clear that Kuhn (and his predecessors) did much to stimulate the concept that science—not just scientists and their practices, but the *content* of scientific understanding at any time—could and should be studied much like any other arena of human activity.

An early instantiation of that concept was the Strong Programme in the Sociology of Knowledge, instituted in the 1960s within the Science Studies Unit of the University of Edinburgh. The sobriquet "sociology of scientific knowledge" (SSK) has been used to distinguish their approach from earlier (Mertonian) sociology of science; "science studies" has come to be applied more generally to work along these lines. In a 1976 book, co-founder David Bloor summed up the tenets of SSK as "causality, impartiality, symmetry and reflexivity." For our purposes, the middle two are the important ones: examination of scientific knowledge should be "impartial with respect to truth and falsity, rationality or irrationality, success or failure" and "symmetrical in its style of explanation. The same types of cause would explain, say, true and false beliefs" (Bloor [1976] 1991, 7). I read this *not* as a claim that truth and falsity are the same, but rather, as a proposal that we suspend trying to distinguish between them while carrying out these investigations. Those are quite different stances—even if a number of commentators do not discern much of a difference, as we shall see in the following section.

The basic concepts of this approach—to be sure, with considerable variation in details and terminology—soon manifested themselves in a wide range of scholarly disciplines; a useful summary includes among them "history, philosophy, sociology, anthropology, feminist theory, and literary criticism" (Rouse 1993, 2). The particular relevance to L&S is emphasized by the fact that this was the very first article published in the SLSA journal *Configurations*. Rouse called these "Cultural Studies of Scientific Knowledge" and characterized them as "refusing to require distinctive methods or categories to understand scientific *knowledge* as opposed to other cultural formations" (3). "Science and Technology Studies" (STS) and other alternate titles have also been suggested; none of them is completely satisfactory in defining the field(s), but I will continue to use "science studies" (which Rouse does as well, later in his article).

As I noted in Chapter 1, it was Rouse's article that first inspired me to look into science studies—including pieces covering a good portion of the breadth of fields he cited—and I found therein much that both considerably interested and somewhat disconcerted me. A significant number of the studies I read—though by no means all—focused on recent or even ongoing scientific episodes rather than historical ones, which I found attractive, being a scientist and not a historian. I was particularly drawn to a couple of them in which the science studiers "embedded" themselves (to use a term since popularized by combat reporters) within research programs: a biochemistry lab at the Salk Institute (Latour and Woolgar 1986), and an attempt to build a novel type of laser (Collins 1992, 51–78). These examinations, which involved close observations of and interactions with the scientists—even extending to active participation—cast considerable light on how science is actually done. But I was disappointed to see how, at the end, the observers largely left the scientists out of the process of *interpreting* their findings. I thought that some degree of cooperation on that level as well could lead to deeper, more universally valid insights.

More generally, the notion that science is a human activity that should be examined as such struck me—again, as a practicing scientist—as entirely reasonable and valid, and I was open to arguments for a significant contribution of social and cultural factors in steering the course of scientific investigations. Frequently in my reading, though, I found it much less clear just how much significance was being attributed to those factors. The language that was used could often be read as claiming a *decisive* role, a stance with which—if it indeed was intended—I was quite uncomfortable, as I expect most scientists would be.

One well-known—and much-contested—example is a book-length study of the 17th-century debates between Robert Boyle and Thomas Hobbes over the nature of experimentation and empirical evidence

(Shapin and Schaffer, 1985). The authors argued that the outcome of those debates largely shaped the subsequent course of scientific development; other commentators on their work made even stronger assertions, such as:

> It is not the mere logical possibility of an alternative science radically disjoint from our own; rather, what compels is that contemporary science is merely the product of following one of several paths that had been equally open at an earlier point in our own history.
>
> (Fuller 1994, 147)

and "Modern science would not exist as we now know it if these people in the seventeenth and early eighteenth centuries had not established this particular way of handling these distinctions" (Hagendijk 1990, 57). To me, there is *nothing* compelling about such counterfactual arguments. Even if a convincing case can be made that social forces guided the outcome of an early debate (many have found the case considerably less than convincing), that is hardly proof that different forces would have resulted in a different outcome whose consequences would have persisted for centuries.

Another example arose from an encounter, very early in my explorations, with a book intriguingly titled *The Golem* (Collins and Pinch 1993). (We will hear more about golems in Chapter 10.) This book, aimed primarily at a general audience, used case studies of several controversial (at the time, at least) 19th- and 20th-century episodes to demonstrate that scientific knowledge is *not* the straightforward result of a perfectly rational and systematic enterprise, as it is often presented. I found their descriptive accounts interesting and entertaining, but again was brought up short by statements that seemed to me way over the top, particularly the concluding remark:

> Why [scientific] debates are unresolvable, in spite of all this expertise, is what we have tried to show in the descriptive chapters of this book. That is, we have shown that scientists at the research front cannot settle their disagreements through better experimentation, more knowledge, more advanced theories, or clearer thinking.
>
> (144)

One *could* read this —perhaps I did, at first—as implying that our traditional understanding of science, consisting of experimentation, interpretation, and theorization, is entirely misplaced, and it is only the social factors—who wins the debates and how—that determine what we accept as scientific fact, a position I could not accept.

I expressed some of my concerns about such apparent overstatement and other matters in an essay (Labinger 1995a); subsequent responses from a number of practitioners of science studies, both in print and in person (especially the opportunity to spend some time with Harry Collins, which arose shortly after I read *The Golem*), convinced me that our positions were actually not so far apart, and could often be resolved simply by more charitable reading—for example, focusing on the key phrase in the quote in the preceding paragraph "at the research front." Thus understood, Collins and Pinch are *not* denying any solid experimental and theoretical grounding of our scientific knowledge, but rather, attempting to show that controversies are not resolved on the basis of those components *alone*, which seems to me a perfectly defensible stance.

I still find their precise phrasing less than optimal: in effect it defines the research front as consisting of unresolved debates, rendering it purely tautological. (Granted, one has to make allowances: everyone—myself definitely included—occasionally succumbs to the temptation to let attractive rhetoric trump exhaustive explanation.) More importantly, I saw little or no concern for how such debates move *off* the research front to become accepted knowledge—what the authors call "textbook science." One philosopher of science remarked:

> The constructionalists … study the first shift of the factory of facts. Quitting work early in the day, they leave us in the lurch with a feeling of absolute contingency. They give little sense of what holds the constructions together beyond the networks of the moment, abetted by human complacency.
>
> (Hacking 1992, 131)

Does science studies thus produce only partial accounts? Yes, certainly; but is that so bad? Another philosopher commented on the treatment in *The Golem* of the notorious "cold fusion" case that ran from the late 1980s to the early 1990s:

> In conclusion, Collins and Pinch and their colleagues have done science studies a great service in choosing to focus on the broader, less methodological issues that naively scientistic accounts of the scientific process have ignored for far too long. But if [they] wish to remain true to their title … they must tell the whole story, not just the parts that have been neglected. Replacing one incomplete story with another serves nobody's interest.
>
> (McKinney 1998, 147)

I am extremely dubious, to say the least, about *anybody's* ability to tell the "whole story" about *anything*. Certainly the relative significance of different perspectives is open to discussion and contest, but a partial story—the only kind we should expect—can be of great value (as McKinney here acknowledges), so long as we take it as *contributing to*, not *replacing*, another story. To echo Dyson's language in a previous chapter, these accounts are complementary, not contradictory.

On the other hand, a number of scientists, as well as more traditional historians and philosophers of science, took a very different view of what they often termed these "postmodern" science studies, denying that they offered *any* such value, and feeling their own work, and science in general, to be under attack. Their responses led to what became known as the Science Wars.

Waging—and Calming—the Science Wars

Skirmishes (in print) between "Defenders of Science" and science studiers began to appear early in the 1990s. While the debate was largely fought *within* academia—primarily by non-scientists therein—it did spread out into the public realm, at least for a short time. An early event in that development was a 1992 book by British embryologist Lewis Wolpert, whose central theme essentially was to *deny* that science is a human endeavor much like any other. Claiming that "science is an unnatural process, quite different from ordinary thinking" (101), he launched a chapter-long attack on the relativism and constructivism he considered to be inherent in the Strong Programme and SSK (101–123). A debate between Wolpert and Harry Collins at a 1994 meeting of the British Association for the Advancement of Science generated enough attention to be reported in the *London Times*—in the Higher Education Supplement, granted, which presumably is read mostly by academic types (Irwin 1994)—while a similarly themed session of the US National Academy of Scholars in the same year, which included participation by notable scientists such as biologist Edward Wilson and Nobel-winning physicist Steven Weinberg, led to an account in the *Boston Globe* (Flint 1994).

That same year saw the full-scale declaration of war, in a book by biologist Paul Gross and mathematician Norman Levitt (1994). They examined several areas of study—constructivist science studies, postmodern literary theory, feminist criticism of science, "radical" environmentalism, AIDS activism, and others—which they termed collectively the Academic Left (probably an unfortunate choice). Assembling exemplars of all the charges mentioned earlier, they argued that their targets in effect constitute a contemporary anti-science brigade. They followed up by organizing a meeting on "The Flight from Science and Reason" at a 1995 meeting of the New York Academy of Sciences (Gross *et al.* 1996), which received considerable public attention.

The most spectacular explosion took place the following year: the so-called Sokal Affair. Physicist Alan Sokal wrote a parody of what he considered to be a typical postmodern treatise on science (Sokal 1996b), and submitted it to the journal *Social Text*, whose editors were putting together a special issue consisting primarily of defenses to the attacks on science studies (Ross 1996), to be published under the title of Science Wars (this was probably the major vehicle whereby the phrase was popularized). Subsequently he revealed that his submission was a test, to see whether it would be recognized as a hoax, and concluded from its acceptance that the journal's editors, and the field in general, lacked the scientific competence to be able to distinguish between sense and non-sense: they were essentially academic frauds (Sokal 1996a). This event totally captured public attention and the popular media: letters on the subject dominated the *New York Times* editorial page for weeks; a major feature article on the Science Wars appeared in *Newsweek* (Begley 1997); and so on. Coverage in journals such as *Science* and *Nature* led to more scientists paying attention—but still not so many. Even at the height of the controversy, when I mentioned to colleagues at scientific meetings and the like that I was working on The Science Wars, by far the most common response was "What's that??"

While the conflict soon faded from public visibility, it continued unabated for some time within academia, represented notably by a volume of essays by (primarily) Defenders of Science (Koertge 1998), a full-book-length statement from Sokal and fellow physicist Jean Bricmont (1998), and a wide variety of further responses published in a wide range of venues (too many to enumerate here). Some of the latter, though, were aimed at least in part at cooling the rhetoric. I participated in two such efforts: a "Science Peace" conference organized by Harry Collins in 1997; and a debate in print edited by Collins and myself (Labinger and Collins 2001). Both featured a mix of science studiers and practicing scientists (including Sokal and Bricmont in the latter venture). It would be far from accurate to say that total comity was reached; but it *is* perhaps fair to say that some of the concerns about science studies have been recognized as overblown and/or misplaced. Since they apply to L&S as well, it is worth spending a little time on that issue here.

The Science Warriors' attacks on the science studies community can be broadly classified into one or more of the following:

- that its members were hostile to science, offering a highly distorted and diminished image thereof.
- that they were encroaching on territory where they didn't belong, having insufficient competence to speak meaningfully about science.
- that they made use of meaningless—or at least impenetrable—language to conceal that lack of understanding.
- above all, that their work—whether intentionally or not—was undermining the acceptance of scientific reliability among the public at large.

While it would take much too much time and space to explore all of these complaints thoroughly, I will discuss some aspects of the first three in the balance of this section, and address what seems to me the most important issue—potential and/or actual harm—in the following one.

With regard to apparent hostility, and intemperate and/or mystical language, there is no shortage of sound bites that can be, and have been, called out. But too often those have been presented in *res ipso loquitur* mode—so egregious on their face that there was no need to try to understand what the authors may have intended. I felt, in contrast, that at least some could be largely defused by the sort of charitable reading I argued for earlier. Take, for example, the "dedication" that Andrew Ross gave to his book *Strange Weather*, which was highlighted by Gross and Levitt (91) as well as others: "This book is dedicated to the science teachers I never had. It could only have been written without them" (Ross 1991). That *seems* uncontestably anti-science—an interpretation I initially endorsed in a review of *Higher Superstition* (Labinger and Weininger 1995)—not to mention rather snarky. But one *could* read Ross as instead suggesting that a traditional scientific education makes it difficult to examine science from a stance outside the tradition. He could have done so without the snark, of course; and one can still argue—as Gross and Levitt do, strongly—that the book as a whole evinces an unacceptable degree of both ignorance of and hostility to science. But then, they apply that characterization to science studies across the board:

> We conclude that hostility to science is, after all, an inextricable element of these postmodern philosophical excursions .... It is mirrored in the remarkable arrogance with which postmodernists address these issues. Virtually all of them claim to discern important intellectual themes and political motifs in past and current science, themes and motifs that are quite invisible to the scientists themselves. These supposed insights rest, as we have seen, on a technical competence so shallow and incomplete as to be analytically worthless. Their arrogance, then, is comparable to that of "creation scientists" in addressing evolutionary biology, or to that of Galileo's persecutors within the Inquisition in their response to his cosmology.
> (Gross and Levitt 1994, 106)

In a similar vein, a 1997 essay on the Science Wars addresses comments made by physicist Steven Weinberg. He characterizes feminist scholar Sandra Harding's description of modern science—"not only sexist, but also racist, classist and culturally coercive"—as questioning the reliability of scientific findings. The essayists offer a counter-description: "Harding's and similar remarks are open to interpretation not as espousals of relativism about the truth of the findings of science, but as claims about the effectiveness of power and prejudice in defining the agendas and applications of science" (Jardine and Frasca-Spada 1997, 226). In both of these examples we see bellicose rhetoric aimed at statements that seem so "obviously" wrongheaded that they can simply be quoted, requiring no interpretation; but we also see how readily one may step back from conflict simply by reading more charitably.

Furthermore, apparently objectionable quotes are sometimes distorted by taking them out of context. For example: " 'the natural world has a small or non-existent role in the construction of scientific knowledge,' as the prominent sociologist of science Harry Collins claims" (Sokal 1998, 11); or "New Cynics like Harry Collins assure us that 'the natural world has a small or non-existent role in the construction of scientific knowledge'" (Haack 2003, 21); or yet again (in an ostensibly even-handed book written jointly by a scientist and a literary scholar): "Collins, for example, made the astonishing remark that 'The natural world has a small or non-existent role in the construction of scientific knowledge'" (Cartwright and Baker 2005, 285). All of these "quotes" cite the same original work (Collins 1981b, 3); but if we go to the source, we find that the *correct* language reads: "One school, however, inspired in particular by Wittgenstein and more lately by the phenomenologists and ethnomethodologists, embraces an explicit relativism in which the natural world has a small or non-existent role in the construction of scientific knowledge." Another version elsewhere (Collins 1981a, 54), which reads "The appropriate attitude for conducting this kind of enquiry is to assume that 'the natural world in no way constrains what is believed to be,'" has likewise been quoted without the introductory qualifying phrase. So criticizing Collins for his "claim" or "assurance" or "astonishing remark" about a negligible role for nature in scientific belief, whether done intentionally or inadvertently, is quite misleading. As actually written, it is a description of a school of thought and/or a methodological prescription for his kind of work, *not* a proclamation of his own dogma.

Another example of what I consider to be bad faith arguing appears in Richard Dawkins's book *Unweaving the Rainbow*, in which he offers the following as an illustration of the confused state of postmodernist thinking:

> The figure/ground distinction prevalent in *Gravity's Rainbow* is also evident in *Vineland*, although in a more self-supporting sense. Thus Derrida uses the term "subsemioticist cultural theory" to denote the role of the reader as poet. Thus, the subject is contextualized into a postcultural capitalist theory that includes language as a paradox.
> (Dawkins 1998, 41)

Just what is Dawkins illustrating here? That each of those three statements is nonsensical? That's far from obvious. Take the first: even though I've read both of those novels (has Dawkins, I wonder?), I don't know whether it might be a valid and insightful comment on Pynchon. But if I ran across it in an essay I would not assume it to be meaningless—just as if I encountered an unclear sentence in an article about some scientific topic in which I lack expertise—but would at least *try* to consider what it might mean within its context. I do agree that the set of three sentences taken together sounds nonsensical, but the passage was *intended* to be nonsense. Dawkins obtained it from a website—the "Postmodernism

Generator"—designed to produce random statements. So here Dawkins is citing a deliberately nonsensical parody as supposed evidence of the nonsensical character of that which is being parodied.

More general disputes over distortion and hostility can be similarly moderated. For example, physicist Robert Park talks about science studies in the context of climate change:

> If scientists all claim to believe in the scientific method, and if they all have access to the same data, how can there be such deep disagreements among them? What separates the two sides in the climate controversy, however, is not so much an argument over the scientific facts, scientific laws, or even the scientific method .... What separates them are profoundly different political and religious world views .... This sort of dispute is seized upon by postmodern critics of science as proof that science is merely a reflection of cultural bias, not a means of reaching objective truth. They portray scientific consensus as scientists voting on the truth. That scientists are influenced by their beliefs is undeniable, but to the frustration of the postmodernists, science is enormously successful.
>
> (Park 2000, 26–27)

Where did that word "frustration" come from? Having acknowledged that world views—among the very things science studiers look at—*do* play a role in the progression of science, Park then claims that those critics are frustrated by science's success, implying that they *must* be anti-science! That seems a rather hefty cultural bias on his own part. A more open-minded commentator might well have reached a different conclusion, something like:

> This sort of dispute is of great interest to those who seek to understand how it can be that while science is to some degree a reflection of cultural bias, and scientists are influenced by their beliefs, nonetheless science is enormously successful.

I believe that to be a much more appropriate characterization of what science studies are after.

A number of years ago I attended a seminar given by my late Caltech colleague Fiona Cowie, on some aspect (I've forgotten what) of philosophy of science, during which she proposed a Central Imperative of Interpretation: do not be too quick to ascribe stupid views to smart people. Much of the ugliness of the Science Wars could have been avoided by judicious application of that principle.

Having (hopefully) moved beyond the question of overt hostility, let us look at the other objections: that by presuming to speak about science without the appropriate credentials, science critics are arrogantly encroaching on science's turf, with (at best) questionable motives, and employing incomprehensible language to do so. At the afore-mentioned British Association debate Wolpert proclaimed that

> not only have [practitioners of science studies] failed to illuminate [the nature of science] but they have actually obfuscated it .... Why are the sociologists of science doing this? I can only give a socio-logical explanation. It's little more than envy.
>
> (Irwin 1994)

Such a stance seems to presume that science is too difficult for non-professionals to be able to offer any meaningful commentary, much as they would like to take on some of what they perceive as the greater stature of scientists. But the converse is rejected: what sociologists and humanists work on is *not* difficult! Their concepts and language should be transparent to anyone, so anything that isn't immediately understand-able must be little more than pretentious nonsense.

In contrast, as I remarked earlier, I believe the *main* reason why the humanities turn their attention to science is neither presumption nor hostility, but rather, the central role of science in our contemporary world. Is it unreasonable that those without professional scientific qualification should want to address that from within their own field (and language) of expertise, and could even have something of value to contribute? Responding (in a personal email) to my review of *Higher Superstition* (Labinger and Weininger 1995), Levitt commented that he was all in favor of non-scientists being *interested* in science, just as he hoped scientists would listen to Mozart. I found that remark telling, because in fact many scientists *play* Mozart, at varying levels of pro-ficiency, from rank amateur (my status, more or less) to virtuoso. At least three of my current or former Caltech colleagues have been pro-fessional concert pianists in addition to their scientific careers. One should not assume that expertise in one realm implies the absence of at least some degree of expertise in another, or that the contributions of those without the highest level of skill cannot be worthwhile on their own terms.

Casualties of War?

Finally, what about the claim of undermining science? As academia was the main battleground, one might first worry about damage therein. I had initially welcomed the possibility that attention to science studies might begin lowering the interdepartmental barriers that characterized much of my university experience; but I was dismayed, as were others (Keller 1995; Labinger 1997), by what was beginning to look like a quite different out-come. There were a number of calls for strong countermeasures, including a proposal from a distinguished chemist that university scientists should challenge colleagues in other disciplines:

Scientists also should confront the sociologists and philosophers at their institutions who are attacking the foundations of science. Presumably, tenure decisions and promotions at universities are based on scholarship, and academic scientists must take an interest in academic decisions in other departments on campus. This is not a question of academic freedom, but of competency.

(Bard 1996)

Even more vituperation can be seen in Gross and Levitt's not-very-veiled threat to excise the humanities from academia altogether:

The humanities, as traditionally understood, are indispensable to our civilization and to the prospects of living a fulfilling life within it. The indispensability of professional academic humanists, on the other hand, is a less certain proposition .... If, taking a fanciful hypothesis, the humanities department at MIT (a bastion, by the way, of left-wing rectitude) were to walk out in a huff, the scientific faculty could ... patch together a humanities curriculum, to be taught by the scientists themselves ... [that] would be, we imagine, no worse than operative .... The notion that scientists and engineers will always accept as axiomatic the competence and indispensability for higher education of humanists and social scientists is altogether too smug.

(1994, 242–243)

and a couple of proclamations offered (by philosophers, not scientists) at the afore-mentioned "flight from reason" conference:

Walk a few steps away from the faculties of science, engineering, medicine, or law, towards the faculty of arts. Here you will meet another world, one where falsities and lies are tolerated, nay manufactured and taught, in industrial quantities .... This fraud has got to be stopped, in the name of intellectual honesty .... Let them do that anywhere else they please, but not in schools; for these are supposed to be places of learning. We should expel the charlatans from the university.

(Bunge 1996, 108)

The sole remedy at our disposal is to quarantine the antiscience brigades and inoculate the rest of the population against them. This requires that those who know something about science—I mean scientists—will have to devote some of their energy to systematic confrontation with the enemies of science.

(B. Gross 1996, 80)

Fortunately the intemperate language died out as the Wars waned, and few, if any, such dire developments came to pass. Indeed, the conflicts were primarily waged *within* the humanities, rather than between humanities and sciences. As I noted earlier, relatively few scientists were even aware of the Science Wars at the time (still fewer are now). I think Sokal's observation (posted on an online discussion site in May 1996) that

> this affair tapped into a pre-existing pool of consternation and resentment among non-postmodernist academics in the humanities and social sciences (of which I, as a scientist, was largely unaware). It's this latter factor that has kept the affair going—in the form of innumerable forums, colloquia and debates—in academia

was pretty accurate. Traditional historians and philosophers of science sensed that their approaches to scholarship were under attack, and fought back. Over time, though, there was some movement towards the position I espoused in the previous section—that their different approaches should be viewed as complementing one another, not conflicting—and relations returned to a state of (mostly) benign neglect, although not entirely without residual resentment and scarring.

For a nice illustration of reconciliation, we can compare case studies of Joseph Weber's attempts to detect gravity waves in the 1970s. (He was unsuccessful; a *vastly* larger-scale program, called LIGO, did finally succeed in 2016.) Sociologist of science Harry Collins and historian of science Allan Franklin took quite different approaches, and ended up with quite different descriptions of the controversy. At the height of the Science Wars, they each defended their own methodology, and criticized the other's, in back-to-back articles (Franklin 1994; Collins 1994). More recently, though, they revisited the case in the form of a *joint* paper that acknowledges the advantage of contributions from multiple approaches. Franklin concluded that:

> For Collins the most significant evidence that he used in his accounts was provided by interviews, whereas Franklin preferred the published record …. The presentation of both views offers a more complete picture of this episode. It is unfortunate that there are not very many pairs of studies of the same episode from these different perspectives. Although we are not suggesting a resumption of the "Science Wars" we do suggest that such conflicting studies can provide us with a better and more complete picture of the practice of science.
>
> (Franklin and Collins 2016, 119)

The deepest concern expressed by Science Warriors was the possibility—the fact, according to some—of damage to the entire scientific enterprise as a consequence of undermining the reliability of scientific knowledge. It is certainly true that in recent years we have seen increasing pressure on funding for and participation in research, as well as resistance to even widely accepted scientific findings. I might also agree that *in principle*, granting a substantial role to social, subjective factors in the determination of those findings *could* provide an effective weapon in the hands of science deniers, even if the practitioners of science studies had no such intention. But has that in fact been the case?

Here are two typical mid–Science Wars statements of that concern. The first was asserted by philosopher Theodore Schick, at a conference on "Science in the Age of (Mis)information" sponsored by the Committee for the Scientific Investigation of Claims of the Paranormal; the second by mathematician Norman Levitt (posted on an online discussion site in May 1998):

> In Feyerabend's view, science is a religion, for it rests on certain dogmas which cannot be rationally justified .... Because most scientists can't justify their methodology, Feyerabend's claims have gone largely unanswered. As a result, Feyerabend's position has become prominent in both academia and the public at large. This has arguably led not only to the rise of pseudoscience and religious fundamentalism, but also to a shrinking pool of scientific jobs and research funds.
>
> (Alexander 1996)

> [T]he fatuities propounded in the name of "science studies" are not, in themselves, particularly dangerous. I can't envision a host of postmodernists taking to the streets in the name of epistemological relativism. Rather, the long-term danger of the phenomena ... is that they help erode such defenses against credulity (and they are very meager) as already exist within the culture. In other words, Paul Forman, Sandra Harding, and company, whether they know it or not, are essentially running interference for P. E. Johnson, Duane Gish, and company.

I find these arguments highly unconvincing, for several reasons. First, it is far from clear that there even *was* a contemporaneous "rise of pseudoscience and religious fundamentalism." A survey carried out by the National Science Foundation at around the same time showed no increase in anti-science attitudes: public support for science remained strong and pretty much constant over the period 1979–1995 (Lawler 1996). The same can be said about the "shrinking pool" of jobs and funding, which in any case always cycle up and down with the economy.

Furthermore, any suggestion of a prominent role for Feyerabend is simply ludicrous. Paul Feyerabend was a well-known iconoclast *within* the ranks of professional philosophers, most (in)famous for proposing that the best descriptor of the scientific method is "anything goes" (Feyerabend 1975); but outside that group he was/is far from a household name. As an experiment, after seeing the report of Schick's talk, I made a point of asking a number of scientific colleagues (mostly chemists, plus a few from other disciplines) their opinion of Feyerabend's impact on science, and got the same response from 100% of them: "Who??" I have no doubt that a survey of the general public, or of Congress and others responsible for determining levels of scientific funding, would have come out just the same.

Levitt cites Paul Forman and Sandra Harding (historian and feminist critic, respectively, both of whom we have encountered earlier) as representative of a "host of postmodernists" who are paving the way for Johnson and Gish, two prominent creationist writers of the time. After seeing his posting I carried out *another* experiment, spending the better part of a (very unpleasant) week looking through their and other pro-creationist works for any evidence of awareness, let alone citation, of science studies, and found nothing. (I did spot a reference to philosopher of science Karl Popper in Johnson's work, but he would hardly count as a postmodernist!)

In retrospect, that shouldn't have been surprising: the *last* thing one would expect from creationists is an argument based upon questioning the possibility of absolute truth! They *do* believe in that; they just don't accept that it can come from science, as Johnson makes clear by attributing the origins of "postmodern irrationalism" to science itself, thus turning Levitt's argument on its head:

> The postmodernist irrationalism that is sweeping our universities is thus the logical outcome of the scientific rationalism that prepared the ground by undermining the metaphysical basis for confidence in objective truth.
>
> (Johnson 1998, 446)

Why would Levitt possibly think otherwise? I suspect it might be something like this: a hypothetical Levitt who reasons in all ways like the actual one *except* for being a creationist might well choose to exploit postmodernist doubts about certainty to undermine acceptance of evolution, paleontology, *etc*. But I seriously doubt whether such a chimerical being exists; or, if one did, that it would have any noticeable influence on the creationist segment of culture.

Nonetheless, these sorts of jeremiads have by no means gone away. Recently psychologist Steven Pinker ascribed more or less equal blame to right-wing politicians and left-wing academics for the subversion of science:

> The highbrow war on science continues to this day, with flak not just from fossil-fuel-funded politicians and religious fundamentalists but also from our most adored intellectuals and in our most august institutions of higher learning .... Students can graduate with only a trifling exposure to science, and what they do learn is often designed to poison them against it .... Many scholars in "science studies" devote their careers to recondite analyses of how the whole

institution is just a pretext for oppression .... More insidious than the ferreting out of ever more cryptic forms of racism and sexism is a demonization campaign that impugns science (together with the rest of the Enlightenment) for crimes that are as old as civilization .... Does the demonization of science in the liberal arts matter? It does, for a number of reasons .... What happens to those who are taught that science is just another narrative like religion and myth, that it lurches from revolution to revolution without making progress, and that it is a rationalization of racism, sexism, and genocide? I've seen the answer: Some of them figure, "If that's what science is, I might as well make money!" Four years later, their brainpower is applied to thinking up algorithms that allow hedge funds to act on financial information a few milliseconds faster, rather than to finding new treatments for Alzheimer's disease or technologies for carbon capture and storage.

(Pinker 2018)

I have already discussed the (lack of) accuracy and fairness of such descriptions of what science studies are about, earlier in this chapter, and will not revisit them here. But the suggestion that young students turn from science to banking because their liberal arts teachers have poisoned their minds, rather than because they recognize how much more money is to be made thereby, strikes me as exceedingly far-fetched. Furthermore, it is simply not true that students are turning away from science, least of all at Pinker's own institution:

A study of Harvard students from 2008 to 2016 found a dramatic shift from the humanities to STEM. The number majoring in history went from 231 to 136; in English, from 236 to 144; and in art history, from sixty-three to thirty-six, while those studying applied math went from 101 to 279; electrical engineering, from none to thirty-nine; and computer science, from eighty-six to 363.

(Massing 2019)

I think it is much more reasonable to hope that attention to science studies in general, and L&S in particular, can contribute to ameliorating undesirable aspects of the trend in academia towards STEM than to proclaim (without evidence) that they provoke the exact opposite.

To be sure, there *are* factions who challenge consensus scientific findings without any inherent ideological commitment such as creationism. (That is not to say that there is no correlation with such commitments, but I will not address that here.) They may be motivated by sincerely held opinions, or financial or political interests, and it perhaps would not be unreasonable to fear that *these* groups could make effective use of "postmodern science studies" to strengthen their positions. But so far as I can see, any evidence that science studies *have* played a causal role in those controversies is at best highly tenuous. In their seminal 2010 work *Merchants of Doubt*, Naomi Oreskes and Erik Conway examine several arenas of science denial, including the adverse effects of smoking, the origins of the "ozone hole," and the climate change debate. Surely, one would think, anyone aiming to promote doubt would cite studies questioning scientific reliability and certainty? But no: in each of these studies they find *scientists*—not many, to be sure, but most or all with respectable credentials—leading the arguments against generally accepted science. (Remarkably, or sadly, or both, they find many of the *same* scientists in more than one of the debates, despite the wide-ranging scientific fields underlying the different cases.) As with creationists, these people do not appeal to science studies to question the *possibility* of reliable knowledge. Instead they challenge the *actual* reliability of claims, although unlike the creationists they employ (to the extent they can) the language and style of science.

Somewhat surprisingly, recent expressions of concern about unintentional support for science denial have come from the science studies community itself. We seem to be moving (inexorably?) towards a "post-truth" society, a development already visible from the work of Oreskes and Conway and related books (Mooney 2005; Otto 2016), and greatly accelerated by the 2016 arrival of Trump and his "alternative facts"–promulgating cohort. Can any of the responsibility for that be attributed to science studies? Bruno Latour worried that "the criticism of science had created a basis for antiscientific thinking and had paved the way for the denial of climate change" (de Vrieze 2017), as spelled out in some detail in an earlier article:

> Do you see why I am worried? I myself have spent some time in the past trying to show "*the lack of scientific certainty*" inherent in the construction of facts. I too made it a "primary issue." But I did not exactly aim at fooling the public by obscuring the certainty of a closed argument—or did I? After all, I have been accused of just that sin. Still, I'd like to believe that, on the contrary, I intended to *emancipate* the public from prematurely naturalized objectified facts. Was I foolishly mistaken? Should I reassure myself by simply saying that bad guys can use any weapon at hand, naturalized facts when it suits them and social construction when it suits them? Should we apologize for having been wrong all along? Or should we rather bring the sword of criticism to criticism itself and do a bit of soul-searching here: what were we really after when we were so intent on showing the social construction of scientific facts?
>
> (Latour 2004, 227)

In a somewhat similar vein (without the soul-searching), Harry Collins and co-authors disagree with the disavowal of responsibility by a fellow sociologist of science (Sismondo 2017):

Sismondo re-packages the history of STS for the post-truth era. His claim is that STS is not to blame for post-truth because the arguments never pointed in that direction. Thus the "science warriors" must have been mistaken because STS had never threatened scientific truth. This distorts the history of our field. The logic of symmetry, and the democratising of science it spawned, invites exactly the scepticism about experts and other elites that now dominates political debate in the US and elsewhere.

(Collins et al. 2017, 580)

Such concerns are understandable but, I think, quite exaggerated. Science studies may well *invite* skepticism and thus provide anti-science forces with a "weapon at hand," but do those forces accept the invitation and actually *wield* the weapon? I saw no suggestion of that in the 2010 Oreskes and Conway book (which was cited by Collins *et al.*), and Oreskes, who continues to be a leading scholar in this area (see, *e.g.*, Supran and Oreskes 2017), told me in a personal email message (December 2017) that she still fails to see any significant role:

I do not know of any evidence that climate change deniers (or others of their ilk) were reading Sandra Harding .... I would say that science studies of the Latourian type was part of a broader cultural questioning of conventional notions of facts and objectivity, that began well before Bruno Latour. But I don't think science studies can be blamed for that: the questioning was valid .... I think that there are broad cultural trends that have encouraged people to question scientific claims, and other forms of authority, and some of this questioning is legitimate. Science studies/history of science legitimately showed that many of these questions can't just be brushed under the rug. We cannot just say "trust us, we're experts." However, I would argue that what *our* work shows (by this, I mean me and Erik Conway) is that the primary culprits in the origins story here is the tobacco industry, who developed the techniques of doubt-mongering about *specific* factual claims and raised them to a high art, and then the other industries, including fossil fuels, pesticide manufacturers, etc., who have applied them shamelessly. And we do have evidence of links between these groups and creationists. It seems to me that the evidence is overwhelming that this has been far more important than the influence of STS scholars. After all, in our work we documented the *direct* links between the tobacco industry and climate change denial. We could have done the same for the chemical industry. But we found no evidence of climate contrarians using the insights of feminist science studies scholars. Believe me, if we had, I would have made hay of that!

Thus, I believe, any fear that the messages of science studies—whether or not they have been correctly understood—have caused collateral damage has been largely unfounded. And Latour and Collins *et al.* are *not* writing to apologize for past sins (as I understand them), but rather, to work out the positive contributions that they and their colleagues can make, as we try to move forward from the strange and worrisome environment we have somehow gotten ourselves into.

My intent in this chapter has been twofold. First, to try to dispel any perception that those who operate at the boundaries between science and the humanities—including but not limited to L&S—might pose any threat to the scientific enterprise or the public reception thereof. Second, and perhaps more importantly, to argue that mutual understanding and acceptance can provide multiple perspectives that deepen our understanding of both enterprises. I would highlight in particular the resolution of the Collins–Franklin debate, mentioned earlier. While their recommendation for studying science from different perspectives was perhaps meant to refer specifically to their particular methodologies,[1] I take it in a more general sense as an endorsement of the entire project represented by this book. In the next chapter, I turn to a more detailed consideration of that project: how I approach L&S.

Note

1 Franklin and Collins cite a couple of episodes where such complementary studies have been carried out independently (while acknowledging that a substantial body of non-published evidence will often not be accessible, especially for anything not quite recent). A few years ago I attempted to do such a project by myself, using *both* extensive published and oral materials to examine a case that was fairly controversial within my own professional circle—inorganic chemists—but hardly noticed by anyone else, that of so-called Bond-Stretch Isomerism (Labinger 2002). I don't have space here to discuss this episode or the conclusions I reached, but I entirely concur with Franklin and Collins: while one *can* tell an interesting and consistent story from either perspective by itself, the combination produces something considerably richer.

4 Models of Engagement

In the foregoing chapters I have tried to establish that exploring relationships between science and literature is a worthwhile endeavor that offers no threat to either side; now, how should we go about doing so? As I noted earlier, many theory-oriented approaches have been promulgated, including a couple that are based significantly on concepts arising from scientific disciplines (I will address those towards the end of this chapter). While such work can certainly make valuable contributions (preferably avoiding the "totalizing" perspective of being the *only* way to do it!), I prefer to pursue L&S without any pre-imposed theoretical structure or conceit, simply looking for relations and resonances between the two realms and exploring the new insights and directions along which they may lead us.

I will lead up to my approach in terms of three terms or concepts— metaphor, analogy, and connection. Although I believe that to a very large degree the first two signify much the same thing—they are often, though not always, used interchangeably—it is convenient to examine some observations that have been offered about each independently.

Metaphor

There is a *vast* body of work about metaphor, and it would be outside the scope of this book (not to mention impossible) to consider more than a fraction of its aspects. A convenient representation of the scope of scholarship in the field can be found in two collections of essays published by Cambridge (Ortony 1993a; Gibbs 2008); the former includes a useful section on metaphor and science. Here I present just a brief examination of its role in science in general, and L&S in particular.

I begin with my understanding of what metaphor is *not*—or, to be more precise, is *not only*. A description of how science studies skeptics think of metaphor can be seen in what has been called the "non-constructivist" position:

> Metaphors characterize rhetoric, not scientific discourse. They are vague, inessential frills, appropriate for the purposes of politicians

DOI: 10.4324/9781003197188-4

and poets, but not for those of scientists because the goal of science is to furnish an accurate (i.e., literal) description of physical reality.

(Ortony 1993b, 2)

or the "standard" view:

Such a view originates with the notion that there are proper, or literal, meanings we can attribute to a given term and that metaphorical usages involve using a word in a distorted or deviant—that is to say, nonliteral—manner.

(Bono 1990, 62)

That view is well represented in *Higher Superstition* (Gross and Levitt 1994), which refers repeatedly to "metaphor mongering" on the part of various science critics, as emphasized in a contemporaneous review by a prominent philosopher of science:

Gross and Levitt also miss the boat with their discussion of the significance of metaphor. Apparently they think that if you turn to metaphor, you simply give up serious thought. Ignoring the fact that the first word of their title is metaphorical, it seems they are totally ignorant of the great body of sophisticated thought—by philosophers, linguists, physicists and others—on metaphor, and of the growing realization that metaphor is not some old crutch to be thrown away as you will. Human thought is deeply, universally and necessarily metaphorical.

(Ruse 1994, 43)

Ruse's last line begins to point at what I consider to be the correct way to think of metaphor; but before going there, let's look at a less dismissive but still overly limited stance. In their (mostly derisive) commentary on a number of "postmodern intellectuals" (all French, as it happens), Sokal and Bricmont offer the following observation:

Some people will no doubt think that we are interpreting these authors too literally and that the passages we quote should be read as metaphors rather than as precise logical arguments. Indeed, in certain cases the "science" *is* undoubtedly intended metaphorically, but what is the purpose of these metaphors? After all, a metaphor is usually employed to clarify an unfamiliar concept by relating it to a more familiar one, not the reverse.

(1998, 10)

I suppose that's a reasonable statement about *one* way metaphor can be used, but it represents a quite impoverished understanding: they portray metaphor as simply an explanatory device, and (worse) one that should only function *unidirectionally*. But metaphor *is* and *does* much more than that: it is *not* just a tool in the hands of authors, but is, rather, an intimate part of *all* speech, writing, and thought, which does not need to be expressly designed to achieve its purpose, as Thomas Kuhn noted:

> However metaphor functions, it nether presupposes nor supplies a list of the respects in which the subjects juxtaposed by metaphor are similar. On the contrary ... it is sometimes (perhaps always) revealing to view metaphor as creating or calling forth the similarities upon which its function depends.
>
> (Kuhn 1993)

Equally importantly, by establishing a relationship between two concepts, it can work to generate new understanding in *both* directions, not just one. Novelist and essayist Arthur Koestler captured this bidirectionality well—even coming up with a neologism (which has not gained any traction, to my knowledge)—in the course of discussing the power of juxtaposing ideas from different contexts. The quote comes from an encyclopedia entry on humor, but I believe it is relevant much more broadly:

> It is the sudden clash between these two mutually exclusive codes of rules—or associative contexts—that produces the comic effect. It compels the listener to perceive the situation in two self-consistent but incompatible frames of reference at the same time; his mind has to operate simultaneously on two different wavelengths. While this unusual condition lasts, the event is not only, as is normally the case, associated with a single frame of reference but "bisociated" with two. The word *bisociation* was coined by the present writer to make a distinction between the routines of disciplined thinking within a single universe of discourse—on a single plane, as it were—and the creative types of mental activity that always operate on more than one plane.
>
> (Koestler n.d.)

Many others have commented similarly on this phenomenon. Chemist Ted Brown introduces his important book on the role of metaphor in scientific work by recalling his experience as founding director of the Beckman Institute at the University of Illinois (a sister organization to the Beckman Institute at Caltech, my home for over 30 years):

> The institute was designed to promote interactions ... to break through the traditional barriers to collaborative work across departmental boundaries ... [it] has succeeded beyond everyone's ambitious expectations .... I soon realized that exchange of significantly different views of the world between researchers from diverse disciplines is an important factor. The different views of a particular problem arise from the prevailing metaphors held by each discipline.

Sharing different metaphorical representations of a problem appears to open up possibilities for creative thinking.

(Brown 2003, *x*)

and physicist-turned-philosopher of science (not the playwright) Arthur Miller offers a similar thought:

An important aspect of scientific creativity is the scientist's ability to create something new by relating it to something already done or understood. Clearly metaphors fit in well here because this is their task.

(Miller 2000, 147)

The reference to creativity in *all* of the last three quotes is no coincidence. A journalist/author's book on metaphor likewise highlights the associative function:

Metaphor systematically disorganizes the common sense of things—jumbling together the abstract with the concrete, the physical with the psychological, the like with the unlike—and reorganizes it into uncommon combinations .... A metaphor juxtaposes two different things and then skews our point of view so unexpected similarities emerge.

(Geary 2012, 2, 9)

while offering support for his view from a particularly notable scientific figure:

[Einstein said] "The physical entities which seem to serve as elements in thought ... can be 'voluntarily' reproduced and combined ... this combinatory play seems to be the essential feature in productive thought." For Einstein, "combinatory play" was the essence of creative thought.

(Geary 2012, 170)

A lengthy account of how such thinking was indeed essential to Einstein's career and contributions (where it is cast in terms of analogy, not metaphor; but, as we shall see, I do not consider this to be an important difference) may be found in Hofstadter and Sander (2013, 452–502).

Commentators from a more literary background exhibit much the same understanding of metaphor. Novelist William Gass, in an essay (partly) about Malcolm Lowry's *Under the Volcano*, observed:

It is far from customary to think of metaphor as a kind of model making ... it is, in fact, tactless to suggest any similarities with science, for isn't it the cold destroyer of the qualitative world, an enemy of feeling, concubine to the computer? More metaphors—and surely false ones .... Metaphors which are deeply committed, which really mean what they say, are systematic—the whole net of relationships matters .... We are inclined to think that in metaphors only one term is figurative ... but this inclination should be resisted .... The terms are inspecting one another—they interact—the figure is drawn both ways.

(Gass 1970, 65–68)

Max Black, recognized as one of the leading authorities on metaphor—and Gass's erstwhile mentor (Monti 2006, 120)—expressed a similar idea thus:

I intend to defend the implausible contention that a metaphorical statement can sometimes generate new knowledge and insight by *changing* relationships between the things designated .... For it may be held that such metaphors reveal connections without *making* them.

(Black 1993, 35)

And a review of a recent quantum physics–imbued novel (Freudenberger 2019) eloquently—and metaphorically!—extols the potency of science-based metaphor in literature:

In the 19th century, Samuel Taylor Coleridge attended public chemistry lectures to expand his "stock of metaphors." Science, he wrote, "being necessarily performed with *the passion of Hope*, it was poetical." In yoking poetry to cutting-edge science, Coleridge was hardly unique: In the 17th century, Milton used Galileo's telescope as a metaphor in "Paradise Lost"; Donne incorporated both the Copernican and the Ptolemaic systems into his verse; Margaret Cavendish wrote about space travel and atoms. Such images, borrowed from science, send us through the looking glass. They cause the universe to expand and contract; they force us to know ourselves in new and startling contexts.

(Hall 2019, 1)

Given all this attention to metaphor from both scientists and "literaturists,"[1] it is unsurprising that the subject has been a major focus of L&S scholarship. In fact I would venture to predict (although I have not done any sort of statistical examination) that L&S studies of any length with *no* mention of the term will be found to be in the minority. Several reference works that self-identify as "Encyclopedias of" or "Companions to" L&S (Gossin 2002; Clarke and Rossini 2011; Meyer 2018a), while not including any discrete pieces on metaphor, index dozens of citations in their contents. A full book-length study is centered on metaphor as the key to relating Einstein and his contemporaries with the literature of the time (Whitworth 2001). Two collections of L&S essays (Peterfreund 1990; Slade and Lee 1990) contain articles that highlight metaphor in their titles (Bono 1990; Hayles 1990b; Zency 1990); the latter collection includes an entire section devoted to metaphor and physics (Schachterle 1990). Obviously any sort of comprehensive survey is impossible here, but I do want to highlight a few points that seem to me particularly illuminating.

Whitworth introduces his book by posing what appears to be an obstacle to work in L&S:

> The study of literature and science raises an ontological problem: faced with two terms which are commonly understood as antithetical, we must explain in what sense we are comparing like with like; the problem appears particularly acute when we are examining the relation of literature to a mathematical science such as physics.
>
> (Whitworth 2001, 1)

and offers a detailed examination of metaphor as a means for surmounting it. He approaches the question of what metaphor *is* a bit hesitantly: he finds a description drawn from the well-known work of Lakoff and Johnson (1980)—"metaphor as the definition of the abstract in terms of the concrete"—unsuitable, preferring the formulation "the unfamiliar in terms of the familiar" (10). That does not entirely satisfy him either—both versions imply a good deal of directionality—but, as he acknowledges, many metaphoric usages are "reversible" (we earlier used the term "bidirectional"); furthermore, assigning degrees of relative familiarity may not be straightforward:

> Nevertheless, the reversibility of the less deeply rooted metaphors allows us to understand how metaphors can circulate between science, literature, and other areas of culture. Not only can scientists understand new phenomena in terms of familiar material objects and social institutions, but non-scientists can defamiliarize the familiar, reconceiving it in metaphors provided by new scientific theories.
>
> (12)

In addition to extended discussions of "highbrow" modernist authors (Conrad, Woolf, Eliot, Lawrence, *etc.*), Whitworth devotes some of his analysis to science writing of the period found in journals aimed at a general audience, commenting:

> The process of juxtaposition in a generalist journal disrupts the agreements that govern the interpretation of articles on history, literature, science, and the arts; furthermore, it implies a programme of comparing ideas between dissimilar disciplines.
>
> (24)

I read these last two quotes as supporting my assertion that the act of associating things drawn from different realms—whether on the macroscopic (articles from different fields in the same journal issue) or microscopic (metaphorically combining diverse concepts) scale—has great potential power for expanding understanding on both sides.

Bono introduces his position by contrasting several examples of what he calls "revisionist" views of metaphor with the "standard" view (a strategy I used at the beginning of this section). Of particular significance is his critique of an essay by Richard Boyd (1993; Bono's citation is to the first edition of this collection), which Bono finds "important" but ultimately inadequate. According to Boyd, "literary" and "scientific" metaphors are quite distinct: the former are created by individual authors as their "property" and remain "open-ended"; complete literal explication (whether by the author or by subsequent critics) is not attainable. The latter become the property of the entire community, by which they are "articulated and developed" to fit our understanding of the empirical world (Bono 1990, 63–64). Using terms we saw in Chapter 1, Bono reads Boyd as making scientific metaphor into a mechanism for "nailing words to things" and "carving nature at its joints":

> This view of scientific metaphors limits sharply the degree to which language can be said to shape scientific discourse. Metaphors are consciously chosen or "introduced" by scientific communities .... Rather than the world's accommodating itself to language, Boyd regards scientific metaphors as part of the process whereby language accommodates itself to the world so that, in the end, "our linguistic categories 'cut the world at its joints.'"
>
> (Bono 1990, 65)

But Bono is very dubious about such an account, especially the idea that science can "control" its metaphors and thereby recruit them into literal explanations of the world (67). He prefers to look at metaphor in science as a "medium of exchange" that can function both to "fix meanings and to disrupt, generate, and transform them" (73). He distinguishes between "intrascientific" metaphors—those that originate within one or more scientific disciplines, such as analogies between the brain and a computer, or between the heart and a machine (73–75)—and "extrascientific" exchange, to which he ascribes a strong role in the evolution of science (it should be noted that Bono is a historian by profession), concluding:

> To the extent, then, that this theoretical analysis implicates scientific discourse in the languages of other, extrascientific discourses, science can no longer regard itself as separate from literature, nor as in complete control of its metaphors and analogies.
>
> (82)

In this view metaphor works interactively, by connecting ideas from different realms (note particularly the juxtaposition of *science* and *literature* in the last quote), to help generate new knowledge, understanding, ways of thinking, about *both*.

Of course, metaphoric usage is not risk-free: there is always the possibility of misleading, as Hesse observed: "A genuine metaphor is also capable of communicating something other than was intended and hence of being *mis*understood" (1966, 164). We can perhaps see this most clearly when indiscriminate appeal to models—which can be considered as elaborate or extended metaphors (the titles of Hesse's and Black's books are, respectively, *Models and Analogies in Science* and *Models and Metaphors*)—has undesirable consequences. There may be what I call a simulacrum effect (after Baudrillard 1994), where a model displaces, or is taken as more real than, that which is being modeled. Turkle (2009) describes two such examples arising from doing molecular modeling on computers:

> Students noted that Peakfinder opened up chemistry to visual intuition .... "We've always been told how molecules are moving, but it was the first time we actually saw what happens." These students' use of the word *actually* is telling. From the earliest days of Athena [a project introducing personal computers to the MIT undergraduate experience] we have seen the paradox that simulation often made people feel most in touch with the real.
>
> (27)

> Physics faculty were concerned that students who understood the theoretical difference between representation and reality lost that clarity when faced with compelling screen graphics.
>
> (31)

> Their program produced a result, and Griffin [a biologist] describes them as "proud of themselves," for "they had gotten this fabulous low-energy structure." But when Griffin checked their result against her understanding of proteins, she realized her colleagues were suggesting a molecule that could not exist. "I tried to explain to them that proteins don't look like that .... I got them books and [showed them] what an alpha helix was and all this stuff and I finally gave up .... Their program told them that this was the lowest energy and they were not going to listen to me."
>
> (68–69)

Another illustration—this one from outside the realm of science—arose from a shopping experience in Paris:

> The girl at the France Télécom store who is asked for a new fax ribbon finds it, places it on the counter beside her—and then spends

fifteen minutes searching through her computer files, her inventory, for some evidence that such ribbons do in fact exist. The ribbon on the counter is an empirical accident; what counts is what is in the system. The reality is the list; the reality is the document ... abstractions pile on abstractions, and by the end you are so distracted that you are unable to face plain facts .... It was not just that you could not see the trees for the forest. It was that you could not see the forest because it was covered by a map.

(Gopnik 2000, 114–115)

All these examples suggest that we must exercise some care in wielding metaphors:

The best way to protect ourselves against the damage of metaphors is to allow the models on which they are based to have as little specific content as possible while still allowing them to serve a constructive purpose. As Arturo Rosenbluth and Norbert Wiener once noted, "The price of metaphor is eternal vigilance."

(Lewontin 2001)

I think "eternal vigilance" is rather too strong, as there is a natural tendency for non-productive or distorting metaphors to atrophy with use over time; but it is always a good idea to be mindful of the chance of introducing unwanted connections.

Analogy

It is not clear to what degree analogy should—or can—be distinguished from metaphor. Indeed, many of the commentators referred to in the previous section are fairly carefree about the terms, often referring to specific illustrations used to discuss metaphor as analogies. One analysis of "the use of analogies and metaphors in science" (Dreistadt 1968) offers several classificatory schemas for analogies: they may be objective or personal, discovered or elaborated (or both), synthetic or analytic. But Dreistadt does not separate out metaphors—the word is used repeatedly *within* all those classes—except at one point, where he defines a metaphor as "an analogy expressed in verbal form" (97). I don't find that very helpful at all. How then would we express *non*-metaphoric analogies? Pictures? Charades?

Another essay, on interactions between the natural and social sciences (Cohen 1993), proposes a fourfold classification based on the nature and degree of similarity, ranging from *identity* at one extreme, through *analogy* and *homology* (referring to systems that are related functionally or structurally, respectively) in the middle, to *metaphor* (which seems to be assigned to cases that are primarily rhetorical in nature) at the other end. Such distinctions may well be useful in their place (that of associating analogy with function and homology with structure comes from biology), but for the present discussion, not so much.[2]

Nonetheless, I do want to come up with at least a loose separation of metaphor from analogy, as some of the works I will look at use only the latter as their topic. I suggest that one may be found in terms of explicit intentionality: analogies have an expressly designed explanatory and/or exploratory function, made clearly visible by the form in which they are presented. We might prefer to call it metaphor when the intended function is not foregrounded; in that sense analogy would appear to constitute a subset of metaphor. But I will take both terms as referring to something that works in much the same way.

Familiar examples of analogies thus employed include the solar system as a model for the atom; waves on water to help understand sound and electromagnetic waves; a branching tree to represent evolutionary development. Many analogies cross over to extra-scientific realms: equilibrium is a central idea in chemistry, physics, physiology, and ecology but also in psychology, economics, international relations, *etc.* Some of these usages may appear to fit, to varying degrees, the unidirectional view of metaphor discussed in the previous section: they aim at using a reasonably well-understood concept or subject area to explain one that is less so. But just as we found that description to be overly restrictive for metaphors, it is likewise so for analogies: they have the capability to expand understanding in both directions:

> Analogy is usually understood ... as the "mapping" of relationships from "source" to "target." As a major example ... the Rutherford-Bohr theory of the atom ... took the known relationship between the sun and orbiting planets and *mapped* it on to the proposed relationship between the positively charged nucleus and the negatively charged electrons. But analogies do not always work this way, particularly, as I will argue, in the most interesting cases.
>
> (Griffiths 2016, 33–35)

Griffiths cites Darwin's analogy between natural selection and domestic breeding as a good illustration of his claim. It is commonly interpreted as trying to explain a new concept—natural selection—in terms of a well-understood one. But Griffiths shows that to be an inadequate account: Darwin argued that domestic selection was also capable of leading to new species, which was *not* generally believed at the time.

> This was an insight he gleaned by testing each half of the analogy against the other, teasing out further features of both. Neither domain of the analogy should be understood as the "base" or "model" for the other: features of each side are adjusted and brought into better alignment to make Darwin's case.
>
> (35)

Griffiths is particularly interested in the role of analogy in contributing to historical understanding (as was Bono in the previous section on metaphor):

> [T]he present study argues that analogy has an extraordinary ability to bring us into contact with other worlds. I understand analogy as the primary formal constituent of *comparative historicism* .... If comparative historicism can be understood as the exploration of how different modes and accounts intersect in time, the study of its constitutive analogies merit (*sic*) special focus, particularly their capacity to formalize new understandings and gain fresh insight into the past.
>
> (28–29)

as well as to L&S scholarship:

> Is analogy simply a literary device? Or is it a natural pattern—a feature of how our world is structured? These questions get to the basic dilemma that constitutes the field of "science and literature" insofar as it hopes to show the value of literary forms in producing scientific knowledge.
>
> (29)

Douglas Hofstadter has produced important work on this topic, much of it set forth in two books whose titles explicitly mention analogy (Hofstadter 1995; Hofstadter and Sander 2013). Both of them are quite lengthy and dense: I do not have space here to do them anything like full justice, and will just draw attention to a couple of points. The earlier book includes a chapter (Hofstadter and McGraw 1995) on creating novel typefaces (both manually and by computer), a long-standing fascination of Hofstadter's (Hofstadter 1995, 402–406). The central issue is: given the key features that characterize a particular font, how should each of the letters of the alphabet be designed so as to best capture those features? At first glance this sounds like a straightforward problem of analogy: what should a given letter in the new font look like, to make it most analogous to its appearance in a familiar font? But a little reflection shows how much more complex a problem is entailed. Just what are those "key" features that define the new font? An equally important (and equally nonobvious) question is: just what *are* the key features that identify a particular letter in any font? So this is in fact a *two-way* exercise in analogy, which can lead us to deeper understanding about the essential nature of the form of a letter or a font. Furthermore, by extending the project to the realm of AI—which involved programs with completely different strategic approaches—they realized a significant insight about *human* cognition:

> In genuine human creation ... any design decision made during the creation of one letter has some chance of influencing the design of all

the other letters in the gridfont …. This inherent unpredictability and instability of what one is doing is precisely the excitement and magic of the genuine creative act.

(Hofstadter and McGraw 1995, 463)

The later book is more explicitly about "analogy as the core of cognition," proposing that "at every moment of our lives, our concepts are selectively triggered by analogies that our brain makes without letup" (Hofstadter and Sander 2013, 3), and arguing against the "standard" view of analogy (and metaphor), using language that echoes the earlier discussion (in Chapter 1) of classification and "nailing words to things":

[M]any of our most familiar categories seem on first glance to have precise and sharp boundaries, and this naïve impression is encouraged by the fact that … every culture constantly, although tacitly, reinforces the impression that words are simply automatic labels that come naturally to the mind and that belong intrinsically to things and entities (14) …. [I]t is misleading to insist on a clear-cut distinction between analogy-making and categorization, since each of them simply makes a connection between two mental entities in order to interpret new situations that we run into by giving us potentially useful points of view on them.

(19)

Hofstadter and Sander close with a passage extolling the power of extended connectivity:

[T]he most advanced breakthrough of Einstein's life came out of an analogical leap that was analogous to another analogical leap—thus an analogy between analogies, or, if you will, a meta-analogy. This recalls a remark once made by the Polish mathematician Stefan Banach: "Good mathematicians see analogies between theorems or theories, but the very best ones see analogies between analogies."

(502)

That idea, along with the phrase in the previous quote—"simply makes a connection"—is central to my approach to L&S, as I will discuss in the next section.

One last point before ending this section. The "standard" view, that we use an analogy between better- and less-understood ideas to help understand the latter, would imply that the more rigorous the analogy—the greater the extent to which it captures "true" relationships between the two components—the better. But with our broader understanding of the term, that need not be the case; indeed, many have observed that a "fuzzy" analogy can be as, or even *more*, productive. Sociologist of science Michael Lynch suggested that extending even an apparently far-fetched analogy (specifically, between the laws of baseball and the laws of physics) to the point where it becomes "absurd" can tell us something deep about words and things (Lynch 2001, 58). In a similar vein, Caltech physicist-turned-biologist Max Delbrück proposed to apply Bohr's ideas about complementarity in quantum physics to molecular biology, recognizing that even though the analogy would almost certainly not hold up, the "paradoxes" that emerged when it failed could be greatly instructive (Roll-Hansen 2000). A commentator on mathematics (G. Polya) exhorted: "[R]emember, do not neglect vague analogies" (Gentner and Jeziorski 1993, 448). Granted, he went on to say "But if you wish them respectable, try to clarify them"—but respectability is not necessarily our main goal!

Devin Griffiths, who began academic life as a molecular biologist, described his (failed) experiment in artificial evolution, and observed that some may think the failure signals "a false analogy between human intent and a mindless natural process. But even if this analogy is false, it is tremendously productive" (Griffiths 2016, 214). Gillian Beer agrees— "Forms of knowledge do not readily merge; they may be askance or cross-grained. But that does not imply failure. Disanalogy can prove to be a powerful heuristic tool" (Beer 1996, 115)—as do two participants in an "interdisciplinary conversation" on poetry and neuroscience: "The usefulness of the false analogy, the train of thought that rattles loose enough to jump tracks, or the leap of connection that misfires but lands somewhere hospitable" (Wilkes and Scott 2016, 337).

In sum, we should look upon analogy—and metaphor—*not* as carefully constructed devices with a well-defined purpose that are worthwhile only to the extent that they succeed in that purpose. Let us, rather, think of them as associations that may have been made intentionally or accidentally, that remain in the intellectual world, and that may be cashed out in unexpected ways.

Connection

Having gone to considerable lengths in the two preceding sections to suggest that "metaphor" and "analogy," in the broadest and most useful sense, can just be viewed as specialized terms for "connection," why do I offer a separate section on the last? Because I want to establish a connection—or analogy—between the way metaphor/analogy works in L&S and the so-called connectionist model for how the brain works. The latter proposes that brain function is basically a matter of forming, and adjusting the strength of, connections between neurons and networks of neurons, such that activation of one or more ("input") leads to a cascading pattern of further activations that constitute the resulting thought or action ("output"). This differs considerably from a model that is computer-like, based on logic and rules.

To be sure, connectionism is not universally accepted, especially an extreme version thereof (which perhaps nobody in fact espouses) that claims it can explain *everything* the brain does—a "blank slate" in which the mind/ brain is completely malleable. Steven Pinker, a strong opponent of that position, argues in his book *The Blank Slate* that such a model cannot possibly account for all of human nature; genetically programmed organizational structuring is required. But he does seem amenable (as I read him) to the idea that connectionism, as broadly described here, does operate, albeit within an architecture that is to some degree pre-ordained (Pinker 2002, 100). To *what degree* it is pre-ordained—the classic nature-*vs.*-nurture argument—is an issue I don't want to get into at all; I will just take it as accepted that such a mechanism is in some ways central to brain function.

For the purpose of this discussion I will focus on one particular instantiation of a connectionist model, that of Gerald Edelman. Edelman began his career in immunology, and shared the 1972 Nobel Prize for explaining how the immune system works. Contrary to earlier postulates, that exposure to foreign substances—antigens—causes antibodies to form in response to the molecular shape of the antigen, Edelman showed that an incredibly diverse population of antibodies is always already present, and responds to an invader by amplifying production of the one(s) that match up with it—a selectionist or adaptational model.

Subsequently Edelman turned his attention to the brain and consciousness, and came up with a somewhat related model for that as well. (Another demonstration of the creative potency of analogy!) The principles have been expounded in a series of technical papers (a layman's summary thereof may be found in Rosenfield 1986) as well as several books written for a general audience (at varying levels of technical complexity). They may be summarized as follows (Edelman and Tononi 2000, 83–86; Edelman 2006, 27–34): connections (synapses) between neurons are established, first during early development—those are substantially influenced by genetics—and then in response to experiences, a process that continues throughout life. Connections between neurons whose activation is correlated with one another are increasingly strengthened, while others are weakened, following Hebb's familiar adage "neurons that fire together, wire together." This results in strongly connected neuronal groups, which interact and synchronize with one another by a process termed "reentry." The connections and interactions that lead to modes of thought and behavior most useful to the organism—the most adaptive ones—are selected for biochemically, by the release of neurotransmitters (noradrenaline and dopamine), in a reward or value system that in turn modulates changes in synaptic strength.

This is thus another selectionist model, which Edelman called the theory of neuronal group selection or, more fancifully (and metaphorically!), "Neural Darwinism." It is not at all computer-like, as Edelman emphasizes:

> I shall take the view that the brain is a selective system more akin in its workings to evolution than to computation or information processing.
> (Edelman 1987, 25)

Such selection ... is not for individual neurons; rather, it is for those groups of neurons whose connectivity and responses are adaptive.
(Edelman 1987, 44)

[B]eing selectional systems, brains operate prima facie not by logic but rather by pattern recognition. This process is *not* precise, as is logic and mathematics. Instead, it trades off specificity and precision, if necessary, to increase its range. It is likely, for example, that early human thought proceeded by metaphor, which, even with the late acquisition of precise means such as logic and mathematical thought, continues to be a major source of imagination and creativity in adult life. The metaphorical capacity of linking disparate entities derives from the associative properties of a recursive degenerate system.
(Edelman 2006, 58)

The last passage explicitly supports my proposal, that the function of neurons in the workings of the brain is closely related to that of metaphor/analogy in human thought.

Edelman links his brain model to an L&S agenda akin to mine, in two successive chapters titled (reminiscent of the Stephen J. Gould title we saw in Chapter 1) "Forms of Knowledge: The Divorce Between Science and the Humanities" and "Repairing the Rift" (Edelman 2006, 68–87). Similar support can also be found from the literature side; for example, an English professor argues for a strong connection between the functions of literature and the brain, based on his extensive study of neuroscience:

My central argument is that literature plays with the brain through experiences of harmony and dissonance that are fundamental to the neurobiology of mental functioning—basic tensions in the operation of the brain between the drive for pattern, synthesis, and constancy versus the need for flexibility, adaptability, and openness to change. The brain's ability to play in a to-and-fro manner between competing imperatives and mutually exclusive possibilities is a consequence of its structure as a decentered, parallel-processing network ... the brain's ability to form and dissolve assemblies of neurons [establishes] the patterns that through repeated firing become our habitual ways of engaging the world, while also combating their tendency to rigidify and promoting the possibility of new cortical connections.
(Armstrong 2013, *ix–x*)

Another article draws analogies between the operations of pattern recognition at both the levels of brain function and cultural practice (Besser 2017).

Of course, such a model must accommodate *mal*function as well: the consequences of selection may well on occasion be *not* particularly—or at all—adaptive.[3] Edelman discusses this in a chapter titled "Abnormal

States" (2006, 106–124), which primarily considers issues of brain damage and the like, but also gives some attention to neuroses and other such non-optimal modes of behavior. While he doesn't go into much detail here—presumably maladaptations result from connections that provide a positive reward on some level and are thus reinforced, even while leading to undesirable consequences on another level—he again expresses his ideas in terms of metaphor:

> Here we come across problems related to the trade-off between specificity and range in a selectional system. Early thought is largely metaphorical, and given its associative powers it can be useful. But if the tension between metaphors in higher-order consciousness and normative values in a culture is given free rein, then it is perhaps not surprising that a rich variety of emotional states and symbolic displacements leading to symptoms can occur.
>
> (Edelman 2006, 122–123)

The nature of my approach to L&S, and the arguments for it, should be apparent by now. When we see possible connections (relationships, associations, analogies, metaphors—any more or less equivalent term will do) between the two realms, we will explore them with particular interest in their potential adaptivity: how, and to what extent, those connections can lead us to new knowledge/understanding/ways of thinking on either side—preferably *both*.

An obvious question arises: how can we evaluate the success or failure of such a venture? Is some sort of objective standard possible, as suggested by physicist Steven Weinberg (with an implicit appeal to Popper's "falsifiability" standard for what constitutes science)?

> Above all, it is the scientists' experience, of being forced by experimental data or mathematical demonstration to conclude that we have been wrong about something, that gives us a sense of the objective character of our work. How often have Fuller or other science studiers had this refreshing experience?
>
> (Weinberg 1994, 750)

Or are such criteria inappropriate to L&S, as argued by Kate Hayles, a leading practitioner thereof?

> [A] colleague from the philosophy of science ... asked about my research. I launched into an explanation of a book I was completing on connections between chaos theory and contemporary literature. When I concluded, she asked what evidence would disprove my thesis. To her, the question was obvious; to me, it was a revelation, for it implied a perspective no one in literary studies was likely to hold. I realized that before I could answer, we would need to have

several hours of conversation about the presuppositions embedded in the question, their effects on guiding inquiry, and the constraints they imposed on possible answers. Only then could we begin to understand why she thought the question was central and why I thought it was beside the point.

(Hayles 1994, 25)

In my career as a practicing chemist I have had Weinberg's "refreshing experience" more times than I like to contemplate; and I certainly believe that it is possible to be "wrong about something" in all fields of work. But I also agree with Hayles that such "scientistic" tests will rarely be all that useful in L&S.

I do *not* mean to suggest that "anything goes"—that we can put out any idea of a connection for consideration, and expect it to be taken as significant *a priori*. There is always more work to be done, as Hayles herself emphasizes:

> In the past, studies in literature and science have tended to follow a characteristic pattern. First some scientific theory or result is explained; then parallels are drawn (or constructed) between it and literary texts; then the author says in effect Q.E.D., and the paper is finished. In my view, every time this formula is used it should be challenged: What do the parallels signify? How do you explain their existence? What mechanisms do you postulate to account for them? What keeps the selection of some theoretical features and some literary texts from being capricious? What are the presuppositions of the explanations you construct, and how do they connect with what you are trying to explain? None of these questions is easy to answer. Nevertheless, if we are to arrive at a deeper understanding of the connections between literature and science ... it is essential not to gloss over the hard issues.
>
> (Hayles 1991, 19)

On the other hand, as we saw in the earlier discussion of "fuzzy" analogies, making connections can be productive even if those "hard issues" have not been worked through completely. Indeed, I argued, they *never* can be—especially not by any one person, working alone, from a single perspective—there are no complete stories. Others agree:

> [D]ifferent "terministic screens" ... vocabularies and codes that direct the attention in particular ways while deflecting it from others ... are typically unable to perceive their defining ratio of insight and blindness by themselves, within the constraints of their own perspectives. That ratio can emerge, however, when one screen is juxtaposed against another ... and this is one reason why the sometimes frustrating disagreements between incommensurable

perspectives can nevertheless be illuminating even when they are not resolved.

(Armstrong 2013, 7–8)

Interdisciplinary studies do not produce closure. Their stories emphasize not simply the circulation of intact ideas across a larger community but transformation: the transformations undergone when ideas enter other genres or different reading groups, the destabilizing of knowledge once it escapes from the initial group of co-workers, its tendency to mean more and other than could have been foreseen.

(Beer 1996, 115)

For me these two quotes reinforce the "Darwinism" metaphor, borrowed from Edelman, that I would apply to my preferred practice of L&S. The power of connection to produce new knowledge and understanding *does* need more than just the act of pointing at it, as Hayles proclaimed; but the pointer doesn't need to do *all* the work! That requisite effort can evolve in unplanned and unexpected ways, within a larger community—ideally one that transcends disciplinary boundaries—that recognizes and adds to productivity. If that *doesn't* happen, the connection will just fade away, much as a non-adaptive synapse atrophies or is pruned out. This is thus a selectionist approach, which resembles Edelman's model of the brain in that regard. If such a mechanism works on the microscopic scale of the individual brain, why not apply it on the macroscopic scale of the community?

L&S from the Science Side?

I want to emphasize that the analogy to a connectionist brain model in the preceding section is not offered as the basis for any sort of *theoretical* approach to L&S; really, it is much more my attempt at justifying a *non*-theoretical approach. However, as I remarked earlier, there are a couple of schools of L&S that are informed by particular fields of science, and in that sense might be considered to be (science) theory-driven to some degree. I will look at those briefly before concluding this introductory part of the book.

The first is generally called "cognitive literary studies," which has been defined by one of its exponents as "the work of literary critics and theorists vitally interested in cognitive science and neuroscience, and therefore with a good deal to say to one another, whatever their differences." He goes on to characterize the field as centered not around any unifying theory but rather, a "shared stance" as well as a "cognitive commitment." The former is that "the instantiation of minds in brains, bodies, and sociophysical environments matters, in understanding literary reading, writing, and literariness itself" (Richardson 2018, 207), while he takes the latter to mean (I am here paraphrasing his quoting

Lakoff) a commitment to make one's own disciplinary work "accord with what is generally known about the mind and brain" from other disciplines (218). A few representative examples of that approach, taken from a "handbook" of the field (Zunshine 2015c), include how writers have treated "body language" in terms of what psychologists call "theory of mind" (Zunshine 2015b); the cognitive factors that influence our emotional response to literary or cinematic depictions (Hogan 2015); and connections between the "Scientific Revolution" and cognitive studies of analogy and metaphor (Crane 2015).

I must say I don't find much if anything objectionable here; and yet, according to Richardson, a cluster of presentations at the (then still-called) SLS meeting of 1998 (I was not present at that one) was met with "remarkable and quite unexpected hostility" (208). A number of reasons were put forth, but primary among them seems to be the perception/fear of scientism: that once scientific interpretations of literature are allowed into the game, they will inevitably take over. Another scholar recalls giving a presentation on her thesis combining poetry with experimental psychology, and "being told by an eminent critic at a literature conference, that he liked my paper but it would be better if I were to 'leave the science out'" (Charlwood 2018, 303). Similar concerns were expressed in other realms of science studies during the Science Wars—which, as Richardson notes, were raging around the time of that meeting. Sociologist Harry Collins worried that "if natural things are to be given a role in analysts' explanations, if the culture of science is to enter the analysis of science ... then it is scientists who must be given the principal word in these areas" (Collins and Yearley 1992, 382).

In my opinion (indeed, this is one of my central premises in this book!) such concerns are misplaced. (Perhaps Collins would now agree, in light of the collaborative work with Franklin we saw in the previous chapter?) L&S can and must be carried out without any sense of superiority or subservience on either side; no field should come to be viewed as what Richardson calls a "master discipline." That applies to cognitive science, as the introduction to the above-mentioned handbook states: "[G]iven what a messy proposition the human mind/brain is and how little we still know about it, striving toward a grand unified theory of cognition and literature is to engage in mythmaking" (Zunshine 2015a, 1). So long as one does not claim that literary studies without insights from psychology and/or neuroscience are *invalid*—and I don't know of anyone who does—nobody should feel threatened by a movement that favors such an approach.

The other L&S movement that takes its lead from a scientific field—specifically, that of evolutionary biology—is commonly known as "Literary Darwinism." Biologist E. O. Wilson, while not the "founding father" (Joseph Carroll, to whom I will return shortly, has perhaps the best claim to that status), is generally recognized as the guru of this school, dating back to his 1998 book *Consilience: The Unity of Knowledge*. That work, also appearing at the height of the Science Wars, proclaimed the lofty goal of reconciling the antagonists by means of "consilience," a term borrowed—or "hijacked," according to one commentator who questions its aptness here (Pigliucci 2016, 250)—from 19th-century polymath William Whewell (who also coined "scientist") to represent the unification of knowledge and understanding of the disparate realms that would signal its accomplishment:

> There is only one way to unite the great branches of learning and end the culture wars. It is to view the boundary between the scientific and literary cultures not as a territorial line but as a broad and mostly unexplored terrain awaiting cooperative entry from both sides.
>
> (Wilson 1998, 137)

That certainly *sounds* like an attractive, even noble ambition. On further reading *Consilience*, however, one encounters language that sounds much more hegemonistic—scientific conquest rather than cooperation:

> The key to the exchange between them is not hybridization, not some unpleasantly self-conscious form of scientific art or artistic science, but reinvigoration of interpretation with the knowledge of science and its proprietary sense of the future.
>
> (230)

> In the last several decades the natural sciences have expanded to reach the borders of the social sciences and humanities. There the principle of consilient explanation guiding the advance must undergo its severest test.
>
> (72)

> The physical sciences have been relatively easy; the social sciences and humanities will be the ultimate challenge. The only way either to establish or to refute consilience is by methods developed in the natural sciences .... Its surest test will be its effectiveness in the social sciences and humanities.
>
> (9)

Wilson does not conceal his commitment to a scientistic, reductionist viewpoint—although, to be fair, he does acknowledge *some* uncertainty about its superiority:

> The cutting edge of science is reductionism, the breaking apart of nature into its natural constituents .... Its strong form is total consilience, which holds that nature is organized by simple universal laws of physics to which all other laws and principles can eventually be reduced. This transcendental world view is the light and way for

many scientific materialists (I admit to being among them), but it could be wrong.

(58–60)

While Wilson offered a number of ideas that might lead to consilience, and its hopeful fulfillment—"When we have unified enough certain knowledge, we will understand who we are and why we are here" (7)—he did not really set forth a coherent vision of his program in this book. But later it became clear that what he had in mind was basically what has come to be known as Literary Darwinism:

> The naturalistic theorists have at the very least clarified and framed the issue. If they are right and not only human nature but its outmost literary productions can be solidly connected to biological roots, it will be one of the great events of intellectual history. *Science and the humanities united!* .... Confusion is what we have now in the realm of literary criticism. The naturalistic ("Darwinian") literary critics have an unbeatable strategy to replace it .... This conception has the enormous advantage that it can be empirically proved to be either right or wrong or, at worst, unsolvable.
>
> (Wilson 2005, *vii*)

The basic tenets of the approach, and their relation to Wilson's conception of consilience, are well represented in a couple of essay collections (Gottschall and Wilson 2005b; Carroll *et al.* 2016). In the former we are told:

> This book attempts to understand the nature of literature from an evolutionary perspective .... choose any subject relevant to humanity—philosophy, anthropology, psychology, economics, political science, law, even religion—and you will find a rapidly expanding interest in approaching the subject from an evolutionary perspective ... [which] had become part of the normal discourse for each of these subjects and is increasingly proving its worth, not only by delivering specific insights that turn out to be correct but also by providing a *single conceptual framework* for unifying disparate bodies of knowledge.
>
> (Gottschall and Wilson 2005a, *xvii*; my italics)

As with the cognitive approach, I have no problem with the idea that this (or any other) particular perspective can offer valuable insights; but I don't accept that any approach or methodology offers *unique* insights. The phrase "single conceptual framework" has a good deal of the flavor of a "master discipline." Likewise, Joseph Carroll's contribution to the same volume contains the sentence "The primary purpose of literary criticism, as an objective pursuit of true knowledge about its subject, is to identify *the* specific configuration of meaning in any text or set of texts"

(Carroll 2005, 94; my italics). I could take issue with a remarkably large number of points in such a short phrase, but I focus particularly on the implication of the italicized (by me) word "the." The proposition that a single "configuration of meaning" in text could possibly be identified strikes me as hugely presumptuous.

As remarked earlier, Carroll was one of the earliest exponents of Literary Darwinism; his book *Evolution and Literary Theory* appeared in 1995. That same year he gave a presentation titled "An Evolutionary Theory of Literary Figuration" at an SLS meeting—this time I *was* present.[4] I eagerly attended Carroll's talk: it constituted my first exposure to what sounded (from his abstract) like a potentially useful take on L&S— particularly from the more scientific side. It was not uninteresting, but what I found most striking was the extent to which Carroll proffered his ideas as the *only* correct way to approach the field. It was (no surprise) not at all well received by the audience.

To be fair, the degree to which the many pieces in these collections (and elsewhere) claim special status for Literary Darwinism varies considerably; both sets of editors deserve commendation for including contrary—or at least skeptical—views of the field (Crews 2005; Hawks 2016; Pigliucci 2016). I don't want to spend much too time on the reasons why many commentators (definitely including myself) are highly dubious about this entire venture; most generally, it is far from clear what "unification" of all knowledge could even mean, let alone how it could be accomplished.[5] Many question whether even scientific knowledge in itself can be said to be unified in any real sense (see for example Galison and Stump 1996).

Stephen Gould (who had a considerable history of disagreements with Wilson) provides a thorough and detailed critique—an entire chapter in a book whose express goal is to "mend the gap" between science and the humanities (Gould 2003, 189–260)—from which I would like to highlight one point: the entirely *contingent* nature of the evolutionary process.

> [W]e have preferred to think of *Homo sapiens* not only as something special (which I surely do not deny), but also as something ordained, necessary, or, at the very least, predictable …. This mistaken view of ourselves as the predictable outcome of a tendency, rather than as a contingent entity, leads us badly astray …. Because we so dearly wish to view ourselves as something general, if not actually ordained, we tend to imbue the universal properties of our species—especially the cognitive aspects that distinguish us from all other creatures— with the predictable characteristics of scientific generalities …. But if *Homo sapiens* represents more of a contingent and improbable fact of history than the apotheosis of a predictable tendency … all the distinctive human properties that feed the practices of the humanities … fall largely into the domain of contingency, and largely outside the

style of science that might be subject to Wilson's kind of subsumption within the reductionist chain.

(226–227)

A study on the central importance of individual variation in mental experience, in the context of the cognitive approach (Otis 2015), similarly raises questions about the utility of analyses that depend on universalities; so does an article about synesthetic readers—those who perceive letters and words as associated with individual colors—offered as the "First Steps toward a History of Neurodiverse Reading" (Rubery 2020, 337). Indeed, in a recent book about brain disorders, a neuroscientist suggests a new science-centered approach to L&S *based precisely on such variation*, effectively turning the Literary Darwinism program on its head:

> Advances in the biology of mind offer the possibility of a new humanism, one that merges the sciences, which are concerned with the natural world, and the humanities, which are concerned with the meaning of human experience. This new scientific humanism, based in good part on biological insights into *differences* in brain function, will change fundamentally the way we view ourselves and one another .... This, in turn, will lead to new insights into human nature and to a deeper understanding and appreciation of both our shared and our individual humanity.
>
> (Kandel 2018, 6; my italics)

Despite these caveats, I do not question the potential utility of Literary Darwinism; I only object to the exaggerated and unjustified portrayal of its omnipotence. In *some* ways it may be consistent with the approach I described in the previous section: make connections—here between texts and evolutionary science—and see where they lead. But paradoxically, despite its sobriquet, it is *not* a Darwinian program—the metaphor I used to characterize my preferred practice—since it is guided by preconceived ideas to such an extent. It is a sort of top-down rather than bottom-up version of Darwinism; its practitioners have in effect taken the well-known aphorism "Nothing in biology makes sense except in the light of evolution" (Dobzhansky 1973) and replaced "biology" with "literature!" The demurring "foreword from the literary side" to one of the above-cited collections, which places a higher value on differences than similarities, seems to me much more appropriate:

> To my mind, evolutionary criticism is just one among many avenues of legitimate inquiry in our field .... It will be important, of course, for "human nature" critics to bear in mind that consilience across disciplines does not require the surrender of one field to the goals and methodological habits of a more basic one. Literary study is not about laws of mental functioning but about a body of heterogeneous

texts. And the social value placed on the most prized of these texts is associated more with their distinctiveness than with their membership in a lawfully governed class. Thus a science of literary criticism, strictly construed, may be neither desirable nor feasible at all.

(Crews 2005, *xiii–xiv*)

This section completes the general discussion of what I am—and am not—aiming to accomplish. Perhaps my main point here is to emphasize that although I *am* a scientist, I do not (intentionally, at least) bring any explicitly scientific agenda to the project. (I might also stress that I am *not* drawing any sort of equivalence between the two approaches discussed here, a claim that I expect would be strongly disavowed by practitioners of both: see for instance Zunshine 2015a, 2–3.) We can now proceed to the second main part of the book, my collection of (hopefully) illustrative case studies.

Notes

1 I do not claim this as my own neologism: a few examples of its use can be found on various online sites (Wikipedia, Google, Yahoo, *etc.*), many of them explaining why it is *not* acceptable. Although I actually like it better than most of the alternatives (the most frequently offered, "men of letters," is both awkward and sexist), I will henceforth refrain from its use.
2 I ran across a much more elaborate treatment that aims for a qualitative *and* quantitative classification of metaphor, in terms of the difficulty of translating passages between different languages; the author proposed a "basic law" in which "translatability keeps an inverse proportion with the quantity of information manifested by the metaphor" (van den Broeck 1981, 84). While in no way wanting to detract from the value of such fine-grained studies in philosophy, linguistics, *etc.* (and readily acknowledging the relevance of translation to my interests, as will be featured in Chapter 7), I am quite dubious about the possibility of assigning metrics to either side of the proposed proportion, or more generally of any such quasi-mathematical approach to the topic. That's *not* what I consider L&S to be about!
3 I offer a modest theory—and lengthy digression—about a phenomenon I find increasingly (and depressingly) common: the "senior moment," wherein one (me) is unable to come up with a word or name for something that one *knows* is perfectly familiar. I suggest this may be a consequence of the formation of *too many* strong connections over the years, tending to lead one down a wrong path, and effectively blocking the right path to the right answer. This thought was inspired by a late-night cascade of thoughts while lying in bed. For some reason (I've forgotten that as well!) I wanted to remember the name of a Greek island we had visited on a cruise a number of years ago. At first all that came to me was that it started with "S," which immediately led to Salonika, a Greek place starting with "S"; and for a long time (at least half an hour) I was completely unable to get that out of my mind, even though I knew full well it was incorrect. Eventually I managed to escape that trap by recalling the name ended in "-ini"; but all that did was lead me to Sammartini—which I knew was

also wrong (an Italian composer, not a Greek locale)—but I remained stuck there for another long time, until *finally* somehow getting to the right name, Santorini. So in this case several strong connections—which could have been completely valid and useful in the right context—proved highly *non*-adaptive, costing me at least an hour of sleep. Of course, it is at least equally possible that the main cause of senior moments is *not* the accumulation of too many positive connections, but rather, age-related degeneration that slows down one's ability to find the right ones. There is probably neuroscientific research out there that bears on this question, but I have made no attempt to track it down—if the latter explanation turns out to be correct, I prefer not to know.

4 Although I was new to the society, I was induced to serve as program organizer (largely because the meeting was in LA, close to Caltech), which turned out to be a remarkably challenging task, requiring sorting large numbers of proposed talks into sessions, making them as intellectually coherent as possible while avoiding conflicts between both similarly themed sessions and presenters' travel schedules, and adjusting all of it up to the very last minute to accommodate changes and cancellations. Somewhere in the middle of the process I swore never to do anything like it again; so naturally, a few years later, I found myself performing the same role (and more) for a 2002 meeting that I hosted in Pasadena. I wonder what a Darwinist analysis of *that* behavior might say!

5 Indeed, it is far from clear just what "unifying enough certain knowledge" means, or how we could possibly quantify it. When I first read the phrase, the image that immediately leapt to mind (this comment is probably gratuitously unkind, but I can't resist) was the Monty Python skit about the "Society for Putting Things on Top of Other Things."

5   Encoding an Infinite Message

Richard Powers's *The Gold Bug Variations*[1]

Bronowski has defined science as "nothing else than the search to dis-cover unity in the wild variety of nature" (Bronowski 1965, 16). This is perhaps the culmination of a belief whose origin G. S. Rousseau locates in the 17th century:

> Nature herself was finite … not an infinite body of knowledge that man could never hope to understand … rather codified in a vast but nevertheless finite set of laws and relationships that would gradually be revealed to man if he persisted.
>
> (Rousseau 1987, 2)

According to this view the diversity and complexity of the world are to be mastered, confined, and simplified, certainly not emphasized or celebrated.

The central message of Richard Powers's third novel, *The Gold Bug Variations* (henceforth abbreviated *GBV*), is that this conception looks from exactly the wrong direction. The wondrous thing about the world is not that "the wild variety of nature" may be encompassed by a finite set of laws, but rather that such a limited basis set can generate infinite variety. The message is eventually understood by Stuart Ressler, a young biological researcher caught up in the race to break the genetic code in the 1950s, and is misread, in various ways, by most of the other characters; these comprise the main themes of the text's narrative. More strikingly, though, the message is not only presented by the text but structurally embedded within it, as Powers has constructed out of *his* limited basis set an echo of systems that can generate infinite possibility. The crucial struc-tural element in this construction, as foreshadowed by the title, is *code*. (I use this word in its "natural" sense, if such a thing exists; at any rate, I do not mean to allude to any discipline, such as semiotics, for which "code" has a special significance.)

In order to see how a message portraying infinity may be encoded, it is essential to distinguish two different functions of code. The more familiar is the simpler: *substitution*. This function is well represented in Poe's

DOI: 10.4324/9781003197188-5

story *The Gold Bug*, where a cipher has the sole function of concealing a set of instructions, which are usable only after reconversion to clear text. A good description of this aspect of code is provided by Hofstadter: "decoding mechanisms … do not *add* any meaning to the signs or objects which they take as input; they merely *reveal* the intrinsic meaning of those signs or objects" (Hofstadter 1980, 164). Code in this sense has no productive power: there is a one-to-one correspondence between the coded message and its deciphered meaning.

A second and more important function of code may be seen in three of the novel's major motifs. Two of these are the genetic code, the set of rules whereby genetic information stored in DNA is translated into protein synthesis in the cell, and computer programming, where code refers to the set of instructions that the programmer actually writes. In programming, the aspect of substitution is still present in some sense: the code, in whatever programming language used, must be converted into machine language. However, the main function of code is not substitution but rather, *generation*: the code is a set of instructions that brings about actions. It does not merely produce another version of itself; rather, it produces the intended output of the computer program.

The distinction is even more obvious with respect to the genetic code. One can distinguish between the set of rules of correspondence between a specific triplet of bases and a specific amino acid (translation) and the synthesis of an enzyme from information coded in a sequence of DNA (expression). This is of course severely oversimplified: with both computers and genetics, there are many more than two levels of meaning for the concept of decoding. (See, for example, Hofstadter 1980, 290–294 and 531–532.) For our purposes, however, this limited dichotomy will serve. In the substitutional sense, the coded message—a list of the nucleotide bases that make up a gene—yields upon decoding just another list, that of amino acids that make up a protein. In the generative sense, though, the result of decoding the genetic message is the living organism! There is thus a multiplicative aspect of code and decoding that takes us far beyond any one-to-one correspondence: the near-infinite complexity and infinite variability of life are generated from the relatively simple set of molecules and rules that comprise the genetic code.

The third coding motif is Bach's *Goldberg Variations*. Although the connection with code is not at first obvious, Powers presents it as another manifestation of code as generator. The *Goldberg Variations* are based upon a simple bass line, only 32 notes long, that functions as the "[thematic germ] on which the entire piece is built" (Powers 1991, 585. For all subsequent references, page numbers alone are given.) The variations "are all obedient, first-filial offspring of the same parent; while different phenotypes, they carry the same underwriting code" (582). Most significantly, the work illustrates how such a generating code can imply, if not actually produce, infinity:

The canons proceed beyond the octave, start all over again at the ninth, as if to suggest, "We could do this for eons." The *Goldbergs* threaten to expand the modest four-note germ of the thirty-two note Base to the scale of infinite invention, a perpetual calendar.

(583)

Since we are dealing here with a literary, not a scientific, work, we need to consider yet another possible coding motif: the function of language as code. Is language a code? Two authorities appear to be in explicit disagreement on this question. A character in a rather different sort of novel proclaims: "To understand a message is to decode it. Language is a code. *But every decoding is another encoding*" (Lodge 1984, 25). Lacan, on the other hand, argues: "[Le langage] n'est pas un code, il est essentiellement ambigu ... [les codes] en principe évitent les ambiguïtés" (Lacan 1978b, 322). In fact, this disagreement is not about the nature of language. Lacan's statement: "la signification ne renvoie jamais qu'à elle-même, c'est-à-dire à une autre signification" (Lacan 1978a, 262) is in perfect agreement with the proclamation by Morris Zapp (David Lodge's colorful character) that "every decoding is another encoding."

Instead, Lacan's belief that language is not a code stems from implicitly restricting code to its simplest level. Discussing computer programming, he claims: "rien ne sort de la machine que ce que nous en attendons .... Elle s'arrête au point où nous avons fixé qu'elle s'arrêterait" (Lacan 1978b, 352). As we see in *GBV* (and as anyone who has dabbled in programming knows well), once a program reaches even a modest level of complexity, it is all too easy to get something unexpected. The same is true for the other coding motifs in Powers's book: "the music is about how variation might ultimately free itself from the instruction that underwrites, but nowhere anticipates what might come from experience's trial run" (585); and "The young scientist left in this gaunt body was himself a product of the code he'd been after, the code that couldn't keep itself hidden from itself" (113). In the generative sense that Powers emphasizes, code exhibits the same ambiguity and potential to produce unexpected meaning that Lacan attributes to language.

To clarify further the connections between language and code, we may consider Black's "substitution view of metaphor" according to which "Understanding a metaphor is like deciphering a code or unraveling a riddle" (Black 1962, 31–32). This view appears analogous to Hofstadter's description of code cited earlier: a *one-to-one mapping*, a static device with no power to expand the scope of meaning. Black rejects this in favor of an "interaction view of metaphor" and quotes Samuel Johnson: "As to metaphorical expression, that is a great excellence in style, when it is used with propriety, for it gives you two ideas for one" (38–44). Here, language is a *many-to-one mapping* which has operational aspects: it produces new meaning rather than simply substituting one meaning for another, in parallel with the generative functions of code discussed earlier. Lacan agrees:

> Toute espèce d'emploi, en un certain sens, l'est toujours, métaphorique .... La compairaison n'est qu'un développement secondaire de la première émergence à l'être du rapport métaphorique, qui est *infiniment* [my emphasis] plus riche que tout ce que je peux sur l'instant élucider.
>
> (Lacan 1978a, 262)

This is clearly a key point: if one wants to represent the generation of infinite variability within the limits of a necessarily finite book, one should take all possible advantage of the multiplicative power of language as code. Powers explicitly addresses this issue in each of his two earlier works. The passage "A map of one inch to the inch, which cannot be spread without covering the countryside, shows nothing that the place itself does not show as well" (Powers 1985, 339) emphasizes the one-to-one *vs.* many-to-one mapping metaphor. In *Prisoner's Dilemma*, we have the image of a time capsule that is intended to show the future *everything* about the current world: "fitting all America into the tube would take a tube the size of all America. But thanks to the recent invention of microfilm, we can fit into this space the blueprint for something far larger" (Powers 1988, 41). Both of these images reappear in *GBV*; the first is discussed at some length (88), while the second is merely mentioned (168).

Two aspects of Powers's use of language in this sense—to expand the scope of his text far beyond its physical confines—merit more detailed consideration. The first is the pervasive use of puns. One reviewer has complained that the punning is excessive and serves merely as a display of virtuosity (Jones 1991). Unquestionably there is more than a trace of showmanship in examples such as "Anyone can have tea for two, but it takes phage to make T4 tumor" (256) or the Marxist (Chico, not Karl) exchange: "I can at least conceive of an oncogene." "I had an Onco Gene, once." "I remember him! Your Onco Gene and your Anti Body" (451). However, this misses a key point: what is a pun, after all, but a one-to-two (or more) mapping of words onto things signified? A pun such as "Cracking the code is just the tip of the *Goldberg*" (369) summons up the image, applicable to icebergs and the *Goldberg Variations* as well as the coding problem, of vast depths only hinted at by the obvious, visible portion.

Allusion is an even more efficient method of generating multiple meaning. One example is both allusion and pun: "a man's speech should exceed his lapse, else what's a meta for?" (517). The pun, the original quotation on which it is based ("Ah, but a man's reach should exceed his grasp/or what's a heaven for?"), and perhaps even the poem from which it is taken (Browning's "Andrea del Sarto"; another early 16th-century painter is a recurring figure in *GBV*) all offer commentary on the text. To borrow from the computer programming motif, allusion is equivalent to calling a subroutine: the author/programmer has only to name (or quote part of) a work to bring the whole into his text.

Powers utilizes this device extensively. There is a key relationship between Ressler and Margaret, the seven-year-old daughter of a research colleague; at each of their two encounters the child recites a poem. In neither case is the entire poem printed, nor are title or author identified—but it is just the left-out parts that are the most crucial. The first ("Margaret, are you grieving/Over Goldengrove unleaving?") (176) is "Spring and Fall" by Gerard Manley Hopkins; the title refers to the important role of the calendar and seasonal changes throughout the book (*vide infra*), while the uncited last line ("It is Margaret you mourn for") is explicitly paraphrased, but not until nearly 400 pages further on (552). Similarly, only parts of the first two verses of the second poem, by Yeats ("When you are old and grey and full of sleep"), are quoted initially (276); another line ("One man loved the pilgrim soul in you") is voiced by a different lover near the very end of the book (634); while the last few lines of the poem ("how Love fled,/And paced upon the mountains overhead/And hid his face amid a crowd of stars"), which appear the most relevant to Ressler's story, remain uncited (except for a fragment in passing (583)) and implicit.

Of course, these allusions are accessible only to the reader who is familiar with or takes the trouble to look up the originals. Powers's fondness for messages that take the form of puzzles is evident: one line from *The Merchant of Venice* is given in hexadecimal ASCII (437), while Beethoven's 9th is coded numerically (572). A description of the *Goldberg Variations* as "slip[ping] inconceivably downstream from the peaceful thematic trickle of its source Brook" (461) only has full significance to those who know that the (helpfully) capitalized word is a translation of Bach. The effectiveness of this technique lies in forcing the *active* participation of the reader, so that all the associations that arise during the decoding process are brought into play along with the coded and decoded messages. The reader must function as central processing unit in these subroutine calls.

An even more elaborate example runs through the text: Ressler is introduced to the *Goldberg Variations* by way of "a two-year-old recording ... in a debut performance by a ... Canadian" (156). The latter is, of course, Glenn Gould, who is frequently referred to but, again, is never explicitly named. There are several parallels between the careers of the real Gould and the fictional Ressler: born in the same year (1932); preferences for solitary and nocturnal lifestyles; premature deaths at about the same age; and most notably, their withdrawals from public life at the height of their abilities (Friedrich 1989). Readers familiar with patent literature may find this reminiscent of a common phrase: "The entire content of [an earlier patent] is incorporated herein by reference." Here we have an *entire character*, whose story may help to understand

Ressler's actions and motivations, incorporated by reference. Like the previous examples, this is a most economical device for keeping a book dealing with the infinite from reaching infinite length.

Beyond these individual examples, the text is constructed upon and unified by *structural* metaphors. Close parallels between the three coding motifs—the genetic code, the *Goldberg Variations*, computer programming—are repeatedly drawn, so that finally each may be understood to encode for each other. The following are just a few of many examples. The *Goldberg Variations* are "Ressler's best metaphor for the living gene" (579), where "in every canon .... Two copies twist about each other with helical precision" (580). (Parallels between genetic coding and canonic structure have also been drawn in Hofstadter 1980, 525–8.) An attempt to give a co-worker a bonus by manipulating a program, which goes disastrously awry, can also be read as a cautionary tale about the potential pitfalls of genetic engineering—a reading reinforced by the chapter subheading in which the attempt is described: "Trace Mutagen" (458). Ressler, asked why he gave up science, denies that he ever did, and produces a batch of his musical compositions as proof. When further asked if any of the pieces have been performed, he replies, referring to the programming foray he is about to launch to recover from the preceding disaster, "Opus One debuts tomorrow" (609–611). Other parallels—philosophical, numerological—abound throughout.

This equivalence principle may next be seen to extend to the entire text itself. The strongest connections are made between the structure of the text and the *Goldberg Variations*. On the most obvious level, the book is organized into 30 chapters, with an introductory poem titled "Aria" and a brief closing "Aria Da Capo e Fine," just as the Bach piece consists of an aria, 30 variations, and a recapitulation of the aria. Only in a few instances are there specific parallels between the contents of a chapter and the corresponding variation. Chapter 25 deals mostly with "Disaster"; Gould's recorded performance of variation 25 depicts not the drama of a disaster itself but rather, the ensuing anguish, as vividly as any musical piece written before or since. Variation 30 is a "Quodlibet," based on popular tunes of Bach's time; one of them is a song whose first line is "I've been away from you for so long." The Quodlibet immediately precedes the return of the Aria, which has not been heard since the beginning of the piece. Chapter 30 features the reuniting of two lovers who have been separated for a long time, brought about when the Quodlibet is sounded on an ATM.

However, the major structural link between book and Bach is much more subtle than any facile parallel between literary and musical events. The narrative is presented in a literary analog of the form that Bach employs for 9 of the 30 variations. Every third variation, starting with number 3, is a canon, wherein a melodic line or voice is accompanied by itself, with the start of the second voice delayed by one-half, one, or two measures with respect to the first. In most cases the voices are exactly parallel, but in two (variations 12 and 15) the second voice is in *inversion*—all intervals are the opposite (up instead of down) from the first. A third voice, present in all but the last canon, does not imitate the first two but serves to unify them.

GBV similarly has three narrative lines. The first begins with Ressler's arrival at the University of Illinois in 1957 to join a team working on deciphering the genetic code. Ressler becomes triply obsessed: with his project; with a married member of the team, Jeanette Koss; and with music, specifically the *Goldberg Variations*, after he receives a recording as a present from Koss. He develops a promising approach to the scientific problem (alone among his colleagues), and at the same time, believes that Koss will leave her husband for him. However, on the brink of success, Ressler discovers that Koss will remain with her husband and leave the project. He abandons the project and disappears from public view.

The second line starts with the meeting between Jan O'Deigh, a reference librarian in Brooklyn, and Franklin Todd, Ressler's co-worker in a computer data processing firm, in 1983. They are fascinated with Ressler and seek to understand his past; at the same time, they become lovers. The couple breaks up; Todd and Ressler are fired when their manipulation of the computer system comes to light, and both leave New York; and O'Deigh is left to her library. Line three is triggered when O'Deigh receives a message from Todd that Ressler has just died. She quits her job and spends the ensuing year studying genetics, retracing Ressler's path to understanding. Todd comes back to her.

The connection between these plot lines and Bach's canonic structure centers on the two love stories, which are highly if not perfectly imitative. In both cases a woman aged about 30, who already has a mate, meets and is attracted to a man of about 25. The new couples progress very gradually, especially on the physical side, towards new and apparently stable relationships; but both founder, ultimately because of the woman's infertility. Some of the parallels between the lines are inverted, just as in two of the canons. Koss desperately wants children but is sterile; O'Deigh is afraid to have children and has had herself sterilized. Koss leaves Ressler to return to her husband, and never sees Ressler again; O'Deigh does not return to her former lover, and eventually gets back together with Todd.

Thus there are two imitative narrative lines, and a third (O'Deigh's year alone) that serves as unifier. How are they organized to give a canon-like structure? One *could* imagine a quite literal canonic form, with the three lines printed one underneath another as in a musical score; but that would merely be annoying: readers are not equipped to take in several lines of text simultaneously as musical listeners can. (For a not-quite-so-literal verbal rendering of fugal structure, see Hofstadter 1980, 311–336.) Instead, Powers frames the narrative lines within the calendar, which is another key structural element of the text. All three lines start around the same date, just after the beginning of summer. In fact, O'Deigh's meeting with Todd, which initiates line two; her learning of Ressler's death, which starts line three; and her reuniting with Todd, which terminates the narrative, all take place on *exactly* the same date, June 23. All three lines run for approximately one year. However, besides the quarter-century separation in time, there is a consistent temporal displacement *within the year*, in many of the key marker events that define the parallel nature of lines one and two. These events—first meeting, first date, first kiss, first consummation, separation—all take place on dates around two months earlier for O'Deigh and Todd than for Koss and Ressler. Presenting the narrative as a time line (Figure 5.1), with one year displayed above the other, we see the two love stories running parallel but with one shifted two months later, just as the musical score of a canon shows two parallel lines with one starting a measure or so later. Thus, Powers exploits the structure that Bach employs for every third variation, but on a much larger scale: not for the individual chapters that correspond to the canonic variations, but for the construction of the *entire* narrative.

Other musical devices are used as well. Most chapters consist of several subchapters, which mark shifts from one plot line to another. Chapter 22, in contrast, contains but a single subheading ("Alla Breve," the marking

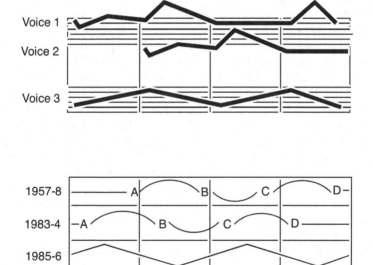

Figure 5.1 Top: Schematic representation of a canon from the Goldberg Variations. Voice 2 imitates voice 1; voice 3 is independent. Bottom: Time line representation of the three narrative lines in *The Gold Bug Variations*. The parallel events in the love stories are represented by letters (A = first date; B = first kiss; C = first consummation; D = breakup). Some of the dates are not given precisely in the text, and may be off by a few weeks in either direction.

of the 22nd *Goldberg Variation*), within which rapid alternation between the three lines takes place. This is the verbal equivalent of a *stretto*, a fugal device where the musical voices follow one another at shorter temporal intervals than the previously established pattern. Conversely, in each of chapters 13 and 14 the subchapters remain within the same narrative line, producing the opposite effect (the musical term is *augmentation*).

Connections between the narrative and the other coding motifs are far less elaborate, though present. The Aria relates the love stories to the structure of DNA: "two couples at arm's length of thirty years bend/in ascending spiral dance around each other" (8), while chapter 19 shows how the narrative can be encoded in the form of a computer program. It really is unnecessary to do any more, though: since the parallels between the motifs are established, connecting the narrative to one of them serves to connect all. Powers has carried metaphor almost to the point of mathematical argument: things equal to the same thing are equal to each other.

One last connection is required: to establish all of human existence as yet another, parallel, infinite coding motif. The key to this connection is the perpetual calendar, which is a recurring theme: it appears as the subheading for the initial Aria (7) as well as in later chapters (164, 265, 626). Like the genetic code, it is a remarkably simple system, "a model of informational economy" (265). But the limited number of possible arrangements of the coding elements—14 counting leap years— nonetheless generates all possible years and, by implication, everything that has happened, or can happen, during those years. The calendar is tied to the text as well, as has already been described: not only are the narrative lines framed by calendar dates, but their termini correspond closely to seasonal changes. "Everything that ever happened happens at equinox" (168). (The significance of the poem "Spring and Fall" is thus strongly reinforced.) Like all codes, the calendar is subject to misreading: O'Deigh is led to a major misinterpretation of Todd's character by incorrectly taking 12/6/85 to mean December 6, rather than June 12 (473); the subheading here is "Transposon." Even mutations are possible, as illustrated by the adoption of the Gregorian calendar—an event whose importance is underscored when librarian O'Deigh selects it for her "Today in History" display (149).

As with all the other coding motifs, the distinction between the simple, repetitive generating principle and its richly varied product is clearly drawn. O'Deigh asks: "February repeats; so does the 3rd: why not the year as well?" (428). Powers replies by demonstrating, over and over, that in a world of infinite possibility, going back and repeating is the equivalent of dying. Ressler tells O'Deigh that he is returning to Illinois to join a new research project; but it is a cancer study that he is joining as subject, not scientist (624). Later, Todd repeats a story Ressler told him, dating from just before their meeting[2]:

[H]e turns the radio on .... It's the Canadian kid .... Playing the piece that woman gave him ... he's shocked to hear that it's not the same piece, not the same performance. It's a radical rethinking .... He can't believe his luck at getting a new recording. But ... the announcer reports that the pianist has suffered a massive cerebral hemorrhage just after releasing this take two.

(636–637)

The last "chapter" of the book runs in its entirety: "Aria Da Capo e Fine. What could be simpler? In rough translation: Once more with feeling" (639). The phrase "what could be simpler?" does indeed repeat the beginning of the book. However, the last phrase must be read as irony: the title, which concludes Bach's work, is properly translated "aria from the beginning *and end.*" The one apparent exception to the pattern is the event that immediately precedes the end, when O'Deigh and Todd resume their relationship. Even if an occasional return is possible, though, stasis is not: when O'Deigh says the relationship would not last because of her sterility, Todd replies:

"And let me ask you another thing." One for the perpetual Question Board. His eyes were full beyond measure. His whole throat shook like a beginner's in wonder at the words he was about to discover. "Who said anything about lasting?"

(638)

If Life itself is to be taken as a coding motif, then it must encode the same message of infinite possibility as the others. Indeed, the various characters' attempts at reading this message—in most cases, not very successfully—comprise a substantial component of the narrative. One member of Ressler's research group, Lovering, sees himself as a subject of, not a participant in, the universe's great experiment of Life: " 'You know what we're going to find out, we researchers? We're going to finally get down to that old secret code in the cell, and the string is going to come out spelling D-U-M-B space S-H-I ...' " (548–549). When he kills himself that evening, he takes all the laboratory animals—his colleagues as experimental subjects—along with him. Koss's eulogy makes it explicit: "Joey lost the signal. Read the message wrong" (551).

Another colleague, Woytowich, walks out on his wife and adored baby daughter because the child appears to be color-blind and he decides, with no other cause for suspicion whatsoever, that she can't be his: "One in several tens of thousands. Which do you think is more likely? A fluke mutation or a woman getting herself plowed?" (564). O'Deigh's error is similar: while doing some research on birth defects for a library user, she becomes so terrified of the possibilities that she has herself sterilized.

> The endless catalog of things that can go wrong ... had killed me ....
> I had assumed that childbearing was a perfected process with a few
> tragic accidents impinging on the periphery. I now saw that the error-
> free lived on a tiny, blessed island of self-delusion.
>
> (385)

Only much later does she realize:

> I've misinterpreted ... from the start .... It's about saying, out loud,
> everything there is, while it's still sayable. The whole, impossibly
> complex goldberg invention of speech, wasted on someone who from
> the first listened only to that string of molecules governing cowardice.
>
> (625)

Todd is also paralyzed by infinity: he cannot write his dissertation on
an obscure Flemish painter because there is always something more to be
learned (30). All have misunderstood the world's message of infinity: with
everything being possible, it *must* be wrong to reject a favorable interpretation even though it appears to be of low probability, or to become
paralyzed because of potential risk. As Ressler notes, "once the experiment gets underway, all possible outcomes are already implied" (193).

From a scientific viewpoint, an interesting aspect of this question of
reading messages from Life is the possibility of thereby gaining insight
into a research problem. Ressler says: "We were looking for the right
analogy, the right *metaphor* that would show us how to conduct the next
round of experiments." (190). That metaphor may direct science is not by
any means a new idea; Bronowski cites Kepler as an illustration:

> Kepler wanted to relate the speeds of the planets to the musical
> intervals. He tried to fit the five regular solids into their orbits. None
> of these likenesses worked, and they have been forgotten; yet they
> have been and they remain the stepping stones of every creative mind.
> Kepler felt for his laws by way of metaphors, he searched mystically
> for likenesses with what he knew in every strange corner of nature.
> And when among these guesses he hit upon his laws, he did not think
> of their numbers as the balancing of a cosmic bank account, but as a
> revelation of the unity in all nature.
>
> (1965, 12)

Hesse discusses the role of metaphor in explaining research, if not necessarily inspiring it, by referring specifically to Black's interaction view
of metaphor (Hesse 1980; also see Weininger 1990, 46–47), discussed
earlier.

Ressler's research group is divided (as was the real scientific world
at the time) between two fundamentally different approaches to solving
the genetic code problem. One was cryptographic in nature: by studying

recurring patterns and their frequencies of appearance, it would be possible to decipher the message. "Pure pattern-breaking attracted the lion's share of fascination" (138). Ressler is at first drawn to this approach: "why dirty one's hands [with experiments] when the problem of pure coding is at stake?" (72) but later comes to realize its futility: "Brute tabulature might work if the underpinning translation were preordained, symmetrical. But there's no guarantee the runaway data enfold formulaic simplicity. In fact, *just the reverse*" (268; italics mine). This takes us back to the primary message: the focus must be on the infinite variations rather than on the simple pattern that engenders them. "[The Code] is just its working out … a figure. A metaphor. The Code exists only as the coded metaphor" (271). This understanding leads Ressler to the correct approach: "There had to be … a way we could get the cell to crack the code for us" (139) and eventually to the design of the successful experimental plan, while most of his colleagues remain mired in dead-end computer-based cipherology.

The final version of Ressler's experiment, which he never carries out, closely resembles the actual work that led to cracking the genetic code. Accounts of the real-world scientists' work, however, do not particularly suggest that they were guided by any such metaphoric understanding. In *GBV*, Ressler recognizes the need to let the cell do the decoding, and works towards an *in vitro* cell-free enzyme synthesis system that could accomplish this. In contrast, in actual events, the *in vitro* system was developed first; the crucial step in the deciphering process, attributed to Nirenberg, was the recognition that the existing technique could be thus applied (Nirenberg 1977, 335; also see Borek 1965, 200–211).[3] Still, it is interesting to speculate about the degree to which extra-scientific world view has influenced, or may yet influence, the course of scientific progress.

Powers sums up Bach's work: "The *Goldbergs* are layered all the way from bottom to top and back down again, with every layer of ordering … contributing to, particularizing, and lost in the next rung of the hierarchy it generates" (583). The same description might be aptly applied to *GBV*. The core message—the overriding importance of the infinite arising from the simple—is embedded in every level: in the metaphor-, allusion-, and pun-rich language; in the individual coding motifs; in the narrative; and in the structure of the entire text, which shows how each of these encodes for the others at the same time that it itself encodes all of them. *GBV* may be seen as a sort of *metacode*: a completely integrated structure that both *contains* and *is* its coded message of infinity. Slade has proposed that:

> As a master coder, the writer of literature knows how to sort through the barrage of information that assaults us daily, to find the messages that may be most valuable but most elusive, and to encode them anew …. To write literature, as the structuralists and their colleagues maintain, *is* to create a world in a text … with proper coding it is

possible to transmit even under noisy conditions a message that is as free from error as the sender cares to make it. That kind of accuracy is a theoretical upper limit, but one any writer can shoot for.

(1990, 13–14)

The last sentence suggests a comment Powers cites about the *Goldberg Variations*: "To compose it, Bach insisted, required only that one work as hard as he did" (586). Whether or not Powers has produced an enduring masterwork, there is little doubt that he followed that prescription.

Notes

1 This chapter is essentially identical to my previously published article (Labinger 1995a); the formatting has been changed to be consistent with the current volume.

2 Ressler's return from his self-imposed solitary existence—including taking up the *Goldberg Variations* again—begins when he meets Todd: "he chose the moment of Todd's arrival to return to the unlistenable piece" (194). One might speculate that the symbolism (Todd = *Tod?*) inspired the choice of name here. Many of the characters' names could be subject to similar readings (Ressler = wrestler?), but in the interest of brevity, this topic will not be pursued any further.

3 As an amusing sidelight, Borek argues earlier in his book (10) that this century will eventually be notable for three things: the monstrous slaughter of its wars, unlocking the energy of the atom, and breaking the Code of Life; it would not be much of a distortion to offer those three as the respective themes of Powers's three novels to date.

6 Is That a Coded Message? It May Not Be So Simple![1]

More on Code and Simplicity

After writing my first piece on L&S (Chapter 5), I recognized that the themes of code and simplicity, both as expressed in *The Gold Bug Variations* and more generally, merit considerably more exploration. The Portland 2007 meeting of SLSA, for which *code* was selected as the central focus, motivated me to delve further into these topics, including organizing a panel on Powers's writings. The present chapter grew out of that occasion.

The ubiquitous role of code in *GBV* was discussed at length in Chapter 5; that of simplicity, somewhat less so, although the catchphrase "What could be simpler?" appears repeatedly. What Powers *appears* to mean by it—at least on first consideration—is that a relatively simple idea or mechanism underlies all the complexity of the world. We see that immediately in the introductory "Aria," which nominally refers to the *Goldberg Variations* but is entitled "The Perpetual Calendar":

> What could be simpler? Four
> scale-steps descend from Do.
> Four such measures carry over
> the course of four phrases, then home.
> At first mere four-ale, the theme swells
> to four seasons, four compass points, four winds,
> forcing forth the four corners of a world
> perfect for getting lost in
>
> (7)

and again on the first page of the first chapter of the book: "Dr. Ressler … laying out all natural history with an ironic shrug: 'What could be simpler?'" (11).

But the word *ironic* is crucial here: Powers wants us to understand that identifying simplicity is *not* simple. A conversation between Ressler and his colleague Tooney Blake reaches a seemingly straightforward conclusion:

DOI: 10.4324/9781003197188-6

"Look," Blake challenges .... "Which do you think will be more complex: a complete, functional description of human physiology, or a complete, functional description of the hereditary blueprint?"

Ressler considers .... As in the old Von Neumann joke,[2] he sees at long last that the answer is obvious. "Physiology is vastly more complex."

"But the more complex is *contained* in the less complex, right? We believe in the simplicity of generating principles."

(363)

that is immediately subverted:

Some equivocation, some sleight of hand here. Can genetics really be said to contain all physiology in embryo? .... Poe's cryptanalyst needed three things to turn the hopeless gold-bug noise back into readable knowledge: context, intention, and appropriate reference. A night of information science has forced Tooney to confront the full width of that triplet.

(363)

In other words, a supposedly simple generating basis—whether referring to the genetic code, wherein each triplet of nucleotides straightforwardly represents an amino acid, or to *The Gold Bug*, where a "simple" substitution cipher is converted to plaintext—is in fact not simple at all. Even just deciphering Poe's cryptic message required a large amount of background. More recently, research has shown that the genetic code operates in ways that are *far* more subtle and complex than originally understood, as we see in the following observations:

35 research teams have analyzed 44 regions of the human genome .... The results ... provide a litany of new insights and drive home how complex our genetic code really is. For example, protein-coding DNA makes up barely 2% of the overall genome, yet 80% of the bases studied showed signs of being expressed.

(Pennisi 2007, 1556)

The more scientists study the genetic code, the more it reads like poetry. In a poem, every word, every line break, even every syllable can carry more than a literal meaning. So too can the molecular letters, syllable, and words of the genetic code carry more biologically relevant meanings than they appear to at first.

(Amato 2007, 38)

Thus, we can be too hasty to take things as simple that in fact are complex, or to see coded messages in what may be something else altogether.

That too was brought home to me while working on *GBV*, whose dedication page consists of the following:

RLS CMW DJP RFP J?O CEP JJN PRG
ZTS MCJ JEH BLM CRR PLC JCM MEP
JNH JDM RBS J?H BJP PJP SCB TLC
KES REP RCP DTH I?H CRB JSB SDG

Obviously—to me—this was in code, and the context of the book immediately suggested some approaches to breaking it: triplets as in the genetic code; groups of eight as in the bass line of the *Goldberg Variations*. As further confirmation (not that I thought I needed any), Ressler successfully interprets a similar message early in the book (47–48), in which the substitution pattern changes from one triplet to the next according to a numeric pattern. So I spent I don't want to remember how many hours, using the notes of the *Goldberg* theme as the basis for varying the rules of encipherment; finally I gave up, and threw myself on the author's mercy. He explained—apologetically, noting that I was far from the only reader who had fallen into the trap[3]—that there was no code at all, or at least not the sort I had taken it for. Each entry save the last was simply a dedicatee's initials (I should have noticed how many of them ended in "P"!), with "?" appearing where he didn't know middle names, and the next-to-last (again, it should have been "obvious") being J. S. Bach. The last stands for *Soli Deo Gloria*, a phrase used often by Bach.

Of course, it is not only authors who present us with ambiguous messages. The same happens regularly in our daily encounters with the world, a consequence of the vast quantity and variability of the sensory input we experience. Richard Feynman expressed how easy it is to mistake randomness for significance with an anecdote:

> You know, the most amazing thing happened to me tonight. I was coming here, on the way to the lecture, and I came in through the parking lot. And you won't believe what happened. I saw a car with the license plate ARW 357! Can you imagine? Of all the millions of license plates in the state, what was the chance that I would see that particular one tonight? Amazing!
>
> (Goodstein 1989, 73)

In this chapter I will first examine what has been said about simplicity as a *desideratum*, whether it is truly so, and how we can be confident of recognizing it. Next I will discuss one of my own chemistry research papers (my first, in fact), and show how my inclination to seek simplicity, and to "decode" a message that wasn't one, helped (mis)lead me to publish a result that was substantially wrong. Finally I will examine a selection of literary examples that illustrate the widespread tendency to over-interpret the universe by taking as coded messages what may actually be just noise.

Is the World Really Simple?

Many have extolled the value of simplicity; some well-known quotes include those of Newton: "Truth is ever to be found in the simplicity, and not the multiplicity and confusion of things"; Feynman: "You can always recognize truth by its beauty and simplicity"; and Thoreau: "Our lives are frittered away by detail. Simplify, simplify."[4] The classic expression of the quest for simplicity is the so-called Ockham's Razor, named for 14th-century English philosopher William of Ockham (or Occam), although he was not the first to propose it. Basically it is a principle of parsimony: given alternate theories that can explain an observation, choose the simplest, the one containing the fewest possible assumptions.

It is important to recognize that Ockham's advice can be taken in two quite distinct ways. Is it a methodological or heuristic principle, a prescription for the best way to proceed with a philosophical or scientific investigation? Or is it a predictive statement about the world, that the simplest solution to a problem is most likely to be the correct one? Such issues have been discussed extensively by philosophers over the years, and I will not go into them in any depth here, except to note that whereas the first interpretation is almost certainly closer to what Ockham and his colleagues had in mind, the latter has taken hold extensively among scientists.

A nice discussion of the Razor, both in general historical terms and as it applies to problems in chemistry, particularly those relating to chemical mechanism, has been provided by three practicing chemists (Hoffmann *et al.* 1996). The mechanism of a chemical reaction is just the sequence of elementary steps (what counts as "elementary" is not always straightforward, but let that go for now) that accounts for the transformation of the starting molecules (reactants) into the products. Mechanism is a topic that has been of great appeal to chemists since around the 1930s, both for its sheer intellectual appeal—solving difficult puzzles is an attractive exercise—and because mechanistic understanding can be a powerful tool for predicting how reactions can be modified to make them go faster, or more cleanly, or to optimize a number of other highly useful attributes. Frequently we find ourselves in the position of having accumulated an array of experimental data that is suggestive, but not conclusive: we have alternatives that cannot be ruled out. Ockham's Razor would thus *appear* to tell us to select the simplest of those.

I have refereed many mechanistic papers over my career, a significant number of which explicitly cited Ockham as justification for their conclusions. Invariably I have recommended, in the strongest possible terms, that any reference to the Razor be excised. Why? Because there are (at least) two serious issues with such an appeal. First, it is not always clear just what constitutes simplicity. Is it a mechanism with fewer steps, or one that avoids a particularly convoluted individual step (sometimes called the principle of least motion), or something else? But (much) more importantly, we have no warranty whatsoever that the world actually *does* work by following the "simplest" path:

> Ockham's razor is perhaps the most widely accepted example of an extraevidential consideration. Many scientists accept and apply the principle in their work, even though it is an entirely metaphysical assumption. There is scant empirical evidence that the world is actually simple or that simple accounts are more likely than complex ones to be true. Our commitment to simplicity is largely an inheritance of 17th-century theology.
>
> (Oreskes *et al.* 1994, 645)

From the purely intellectual point of view, an acceptable application of Ockham might be the following: at any particular point in an investigation, if there is an explanation that accounts reasonably well for the observations in hand, there is no reason to postulate additional complexities unless and until further experimental findings lead that way. (Often, I've noticed, Ockham seems to be cited primarily as an excuse for not doing those additional experiments!) But in terms of any practical implications one hopes to glean from mechanistic study, it is far from clear that it has any validity whatsoever.

As an illustration of the problematic appeal to simplicity, let us briefly examine a case from chemical history: the quest to understand atomic composition. A key insight, generally credited to Dalton in the early 19th century, was that matter is composed of atoms that combine with one another in fixed proportions, such that the relative weight of an atom could be determined for each element. By 1815 around a dozen such weights were known; some of them were incorrect (often by a factor of two, since the formulae of the various molecules used in the determination were not known—a problem that was not finally resolved until nearly 50 years later), but it appeared to be the case that, within accepted experimental uncertainty, all the values were integral multiples of that of hydrogen, the lightest of all. That led English chemist William Prout to hypothesize that all matter is composed of entities that have the atomic weight of hydrogen (he called them "protyles"): a hydrogen atom would be just one protyle, while atoms of heavier elements would consist of the appropriate number somehow fused together (Scerri 2007, 38–42).

Research during the next few decades established weights for additional elements and improved the precision of the measurements. Nearly all came out "close enough" to sustain the hypothesis, but there was one glaring and persistent exception: the best value for the atomic weight for chlorine (which Prout had taken to be 36) was adjusted to 35.5, and there it remained. Indeed, a periodic table published as late as 1902 shows *all* the elements (now numbering around 75) *except* for chlorine with integral atomic weights, even though by then there were many others that had been determined to diverge significantly from whole numbers (Ihde 1964, 252). Here was a clear inconsistency; how to resolve it? Various proposals included that Prout's hypothesis should simply be abandoned (but replaced by what?); that the fundamental building block was not a hydrogen atom of weight 1, but a subatomic particle (of unknown nature) of weight 0.5, or even 0.25; that the experimental result for chlorine was wrong for some unknown reason, such as a contribution of a small amount of "luminiferous ether" or even a need to modify Newton's law of gravitation (!), and so on (Benfey 1952).

If we were to apply Ockham's Razor to this dilemma, what would it tell us? Even if we only want to use it as a heuristic guide for how to proceed, it is far from clear whether, or how, any of these choices is more or less simple than any of the others, except perhaps for the "give up for now" option; and we have an even greater problem if we hope to find the "right" answer. Each of them seems to require either wholesale revision of well-accepted science, or postulating a new entity or concept. Do we have the slightest reason to believe the Razor could point the way forward?

The conundrum was not resolved until the 20th century, and the solution *did* require the introduction of a new entity—the neutron—and a new concept—isotopy. Atomic weight was shown to be a measure of the number of protons and neutrons combined. The number of protons defines the element—chlorine has 17—but most elements exhibit more than one isotope, with different numbers of neutrons. A large fraction of these are either unstable (radioactive) or found in nature in very small amounts; but chlorine just happens to exist in *two* abundant isotopes, with atomic weights 35 and 37, present in nature at around 75% and 25%, respectively. Thus, the measured value is the weighted average, around 35.5. (Note that if we equate the proton with Prout's protyle, he wasn't *entirely* wrong!)

How Do We Know Whether an Experimental Observation Is a Coded Message or Just Noise?

Many more examples in the history of chemistry, or science more generally, could be cited to illustrate the questionable value of any criterion of simplicity; but now I want to examine the case of my own erroneous first paper, which dealt with a chemical reaction mechanism. My error had something to do with a misplaced belief in the power of simplicity—I think I was at least nicked by the Razor—but a problem in "reading code" played a more central role.

While working on the talk mentioned at the beginning of this chapter, I recalled something I had encountered in a seminal work of science studies: the "anthropological" study of the Salk Institute cited in Chapter 3. The authors conceived the notion that the primary activity going on was the production of *inscriptions*, which included the output of scientific instruments—figures, images, and graphical traces; the sub-heading of the section where this concept was introduced is "Literary Inscription" (Latour and Woolgar 1986, 45–53). That recollection led to another resonance for me: virtually all of my research involved generating and interpreting graphical instrumental output. Could it be useful to view such traces as a kind of *coded* message, the understanding of which has something in common with that of a text? I will use that flawed first paper to try to show that is indeed the case. While I believe the experimental methodology would be instructive and (reasonably) comprehensible even to a non-technical audience, it *is* more than a little complex and challenging, so I give here only a sketch, hopefully sufficient to get the main points across. For those interested, a more detailed discussion may be found in the Appendix.

The study had to do with a reaction of a class of molecules known as alkyl halides, which consist of a halogen atom—fluorine, chlorine, bromine, or iodine—which we will label as X, attached to a hydrocarbon group. Such species have been known for centuries; the simplest are the methyl halides, CH_3X. Alkyl halides have a number of important uses; methyl bromide, for example, is used (but highly controversially) as a pesticide. A common reaction pattern known as "nucleophilic substitution" consists of the replacement of the halide by some other group N: a so-called nucleophile (which could be another halide or something else). Frequently the mechanism of such reactions involves a "simple" one-step process, in which the entry of the incoming group and the departure of the leaving group—or, stated somewhat differently, the breaking of the C-X bond and formation of the C-N bond—are essentially simultaneous; this is termed an S_N2 mechanism. The particular subject of my research was a related reaction called "oxidative addition." Here too the carbon-halogen bond is cleaved, but not by a simple substitution; instead, both halves undergo *addition* to a metal center M. These two transformations are represented in Figure 6.1.

As noted earlier, mechanistic investigations in general can have practical as well as intellectual interest, and that was true here as well: the oxidative addition process has a number of practical applications whose utility can often be optimized with mechanistic knowledge (Labinger 2015). We were eager to determine the detailed mechanism, and envisioned two possibilities: one *stepwise*, where an initial S_N2 step forms the metal-carbon bond and a free halide ion that subsequently recombines with the metal center; and the other *concerted*, in which the new M-C and M-X bonds form simultaneously. Chemists have a variety of methods for distinguishing between mechanisms; many of them involve studying how the rate of the reaction depends upon different conditions, such as concentration, temperature, and the solvent in which it is carried out. Those

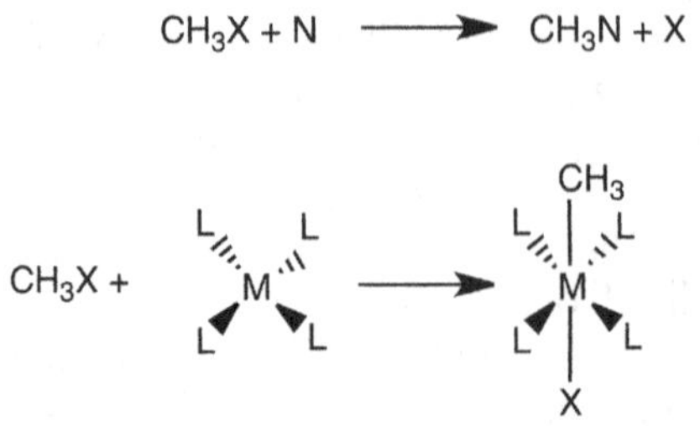

$$CH_3X + N \longrightarrow CH_3N + X$$

Figure 6.1 Representations of nucleophilic displacement (top) and oxidative addition (bottom) reactions of a methyl halide. In the latter, the two halves
of the methyl halide add to *opposite* sides of a species consisting of a
metal center with four attached groups. By convention, bonds drawn
as solid lines lie in the plane of the page, while those with bold and
dashed wedges go in front of and behind the page, respectively.

behavioral trends are very well understood for nucleophilic substitutions;
and earlier experiments had clearly implicated an S_N2 mechanism for oxidative addition of methyl halides (Chock and Halpern 1966). But was
that generalizable to other alkyl halides?

In retrospect, I think we were too inclined to expect that *would* be the
case (which of course we should not have been!), a prejudice that probably had much to do with a subconscious application of Ockham's razor,
since that would be the simplest alternative. Or would it? As discussed
earlier for mechanistic questions in general, it is far from clear *which*
of the alternatives should be considered simpler. If the mechanism were
analogous to the very common S_N2 process, that would require no introduction of a new concept: would it therefore be simpler? Or is it simpler to have a reaction that proceeds in a single step instead of two?
Or is the latter actually *more* complex despite being a single step? The
convoluted atomic motions required to insert the metal center (which
includes additional bulky attached groups) into the alkyl halide and have
the two halves end up on opposite sides of the final product were unpre-
cedented and looked rather unnatural (it was dubbed the "bacon-slicer"
mechanism).

Fortunately, there is another mechanistic test applicable to this question,
generally considered the most conclusive: that of stereochemistry—the
geometrical consequences of the reaction. A more thorough explanation
of the stereochemical aspects of organic chemistry relevant to this study
(and also to a topic to be introduced in Chapter 9) is presented in the
Appendix. Briefly, an alkyl halide has four groups attached to the central
carbon, arranged in the form of a tetrahedron. In an S_N2 reaction the
incoming group approaches from the direction *opposite* to the leaving
group, such that the geometry about the carbon is effectively turned
inside out (called *inversion*). That would be true for oxidative addition

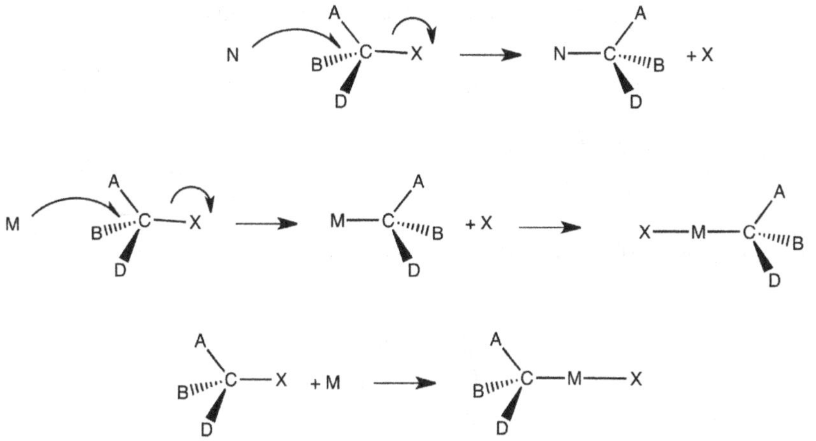

Figure 6.2 Stereochemical consequences of alternate mechanisms. Top: an S_N2 reaction in which nucleophile N displaces leaving group X by approaching carbon center C from the side opposite to X, resulting in inversion. Middle: an oxidative addition proceeding by a two-step mechanism, the first being S_N2-like, again resulting in inversion. Bottom: an oxidative addition proceeding by a single-step mechanism, where the metal center effectively inserts into the C-X bond, resulting in retention. As long as the three groups (A, B, D) attached to the carbon center are all different, the two outcomes can be distinguished. (The other groups attached to M are not shown, for simplicity.)

as well, *if* it followed an S_N2-like mechanism. In contrast, the alternative insertion mechanism, where the metal approaches directly *at* the carbon-halide bond, would leave the geometry unchanged (*retention*). This is shown schematically in Figure 6.2.

This geometric test is not available for methyl halides—the high symmetry of the methyl group makes the two outcomes indistinguishable—but it is for more complex alkyl halides. We designed a first-generation probe molecule for which the answer would be obtainable using a technique known as nuclear magnetic resonance, or NMR. That technique, along with the structure of the probe molecule and how it allows the mechanistic distinction to be made, is briefly explained in the Appendix. Without going into detail here, we could predict with confidence that a reaction with inversion would result in an NMR signal consisting of four broad lines, while retention would give only two broad lines. Figure 6.3 shows, schematically, what we expected for the two alternative cases, and what we actually observed. The result *seemed* to point pretty conclusively to one result—inversion—and we hastened to publish that conclusion (Labinger *et al.* 1970).

That decision turned out to be rash. We weren't entirely comfortable about the evidence: both the relatively poor signal strength (NMR signals

NMR spectroscopic determination of reaction mechanism

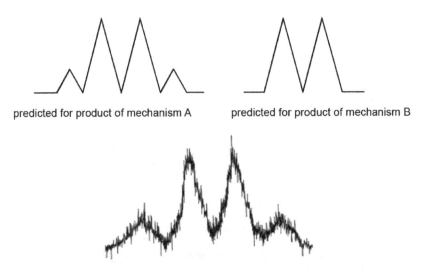

predicted for product of mechanism A   predicted for product of mechanism B

Figure 6.3 Top: Idealized representation of the NMR signal expected for the mechanistic probe reaction, depending upon whether it proceeds with inversion or retention. Bottom: the actual observed spectrum, as published.

tend to be inherently weak; with the technology available in 1970, it was necessary to collect a large number of spectra and "average" them by computer processing to see much of anything at all) and the featureless width of the signals were somewhat concerning. Indeed, we were already working on a new test molecule that we anticipated would solve the latter problem, and we should have waited for the results obtained therefrom. Our failure to do so was probably due in some part to the considerations of simplicity sketched earlier.

When we *did* carry out studies on the second-generation probe, we got a quite contrary result (Bradley *et al.* 1972). That reaction produced *both* possible outcomes —half the product molecules showed inversion, half retention—clearly inconsistent with our first conclusion, and apparently requiring us to postulate a novel mechanism after all. It took us a little while to reconcile these findings, but with the aid of more advanced NMR methodology (just being introduced while the work was going on) we were able to determine that indeed there *were* two different products in our first experiment as well. But their signals overlapped, coincidentally (and unluckily) in just such a way as to produce three of the four broad lines we had originally seen, as shown in Figure 6.4. Whence came the fourth signal in the original spectrum, missing in the new one? That, as we were able to demonstrate to our (and the referees') satisfaction, had

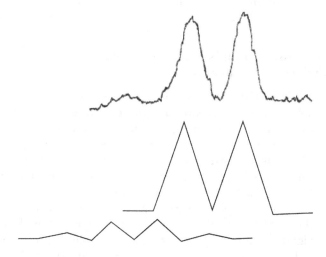

Figure 6.4 Top: NMR signal obtained using an improved spectrometer, showing only three broad peaks, instead of four as in Figure 6.3. Bottom: Idealized representation showing how the top signal results from the accidental overlap of *two* separate signals, a doublet and a much weaker quartet. The right-hand broad peak seen in Figure 6.3 but not here can only have been due to noise in the earlier spectrum.

just been noise, a consequence of the weak signals we were working with, and (again coincidentally) turning up in just the position where we could mistake it for an integral component of a single quartet signal.

Fortunately there was no significant long-term harm caused by the initial error. The entirely novel mechanism implicated by the second probe studies proved to be correct (*pace* Ockham!) and of considerable general importance, serving as the basis for my eventual dissertation and much subsequent study. It took me a long while to derive any general lessons from this episode, but now I see some. It is true that we were deceived by what may appear to be unlikely coincidences; but subsequent similar events have happened often enough in my research career to make me question Einstein's proclamation that "*Raffiniert ist der Herrgott, aber boshaft ist er nicht*" (Pais 1982, *vii*): "God is subtle, but not malicious." Our overconfidence about understanding simplicity—the problem discussed in the first part of this chapter—led us to interpret noise as a coded message. As we shall see in the following section, that mistake is frequently manifested in literature. Perhaps an earlier recognition of what now seem to me clear connections to L&S could have helped us avoid those pitfalls. Is that far-fetched? Maybe not. As chemistry Nobelist Tom Cech commented, "Scientists need the same skills as humanists to cut through misleading observations and arrive at a defensible interpretation, and intellectual cross-training in the humanities exercises the relevant portions of the brain" (1999, 210).

Distinguishing Message from Noise in Some Works of (Mostly South American) Literature

Trying to uncover coded messages in objects and texts—words and things—has a long and extensive history in both literature and real life. In the former, one need only turn to the vast body of detective fiction, including the explicit decoding (deciphering, more precisely) exercise central to *The Gold Bug*. In the latter, the hunt for occult meaning in one particular text—the Bible—is still more vast: a Google search on "hidden messages in the Bible" turned up over 16 million hits.

A well-known literary example appears in *War and Peace*, when Pierre learns from his Masonic brethren about a prophecy concerning Napoleon, resulting from assigning numbers to letters of the French alphabet according to $a = 1$, $b = 2$, and so on up to $i = 9$; then $k = 10$ (j is omitted), $l = 20$ … $z = 160$. In that system "l'empereur Napoleon" yields 666 (but only if an extra 5 for the elided "e" in "l'empereur" is added in), the "number of the beast" in the Book of Revelations. Likewise, "quarante-deux" gives 666, and Napoleon turned 42 in 1812; hence Napoleon's invasion of Russia in 1812 must be apocalyptic. Pierre believes he himself must have a role to play, and tries to apply the numerology to prove that; after a number of unsuccessful attempts he works out that his *own* name can be made to add up to 666 if he calls himself "l'russe Besuhof"—Besuhof the Russian. Of course, that's cheating: his name in Russian (Безухов; in English it's rendered as Bezukhov or Bezuhov) transliterates into French as Besouhoff or Besouhof, certainly not Besuhof, which would give a quite different "u" sound; also, one would never elide the "e" in "le russe." But none of that worries Pierre (Tolstoy [1869] 1968, 801–802):

> The discovery excited him. How, by what means, he was connected with the great event foretold in the Apocalypse he did not know, but he did not for a moment doubt the connection. His love for Natasha, the Antichrist, Napoleon's invasion, the comet, six hundred and sixty-six, and *l'russe Besuhof*—all this had to ripen and burgeon forth to lead him out of that spellbound petty sphere of Moscow habits in which he felt himself held captive, and guide him to some great achievement and great happiness.

A 20th-century author strongly associated with the search for meaning in a cryptic universe is Argentinean Jorge Luis Borges, many of whose short stories highlight that theme. In *Death and the Compass*, for example, the protagonist detective Lönnrot investigates a series of murders by trying to interpret a variety of clues—textual (including a dose of Biblical exegesis), geographic, geometric—that for him all point to one more murder in a certain place, where he goes to apprehend the criminal. Unfortunately his deductions are not quite right, as some of the apparent clues are unrelated to the case, while others are open to multiple alternate readings. He has no reliable way of discriminating significant coded messages from irrelevant noise—a situation that applies to the reader as well, as one commentator points out (Martínez 2012, 44–55). At the story's end, Lönnrot is revealed to be the intended victim, and is duly shot (Borges 1964b). Borges's world, clearly, is governed by neither Sherlockian logic or Ockhamian simplicity. A similar theme forms the basis of Umberto Eco's 1980 novel *The Name of the Rose*, whose conclusion also involves a trap set for the investigator by the murderer—although this time the former escapes and triumphs. Eco explicitly pays tribute to his predecessors, *inter alia*, by the names chosen for the two antagonists: the investigator is William (after Ockham) of Baskerville (after the hound), while the murderer is Jorge of Burgos.

The ultimate expression of unknowable meaning—we might call it an *expando ad absurdum*—may be found in Borges's *The Library of Babel* (1964c), which offers the allegory of a universe of text:

> The universe (which others call the Library) is composed of an indefinite and perhaps infinite number of hexagonal galleries .... Twenty shelves, five long shelves per side, cover all the sides except two .... each shelf contains thirty-five books of uniform format; each book is of four hundred and ten pages; each page, of forty lines, each line, of some eighty letters.
>
> (51–52)

> [A] librarian of genius discover[ed] the fundamental law of the Library. This thinker observed that all the books, no matter how diverse they might be, are made up of the same elements: the space, the period, the comma, the twenty-two letters of the alphabet. He also alleged a fact which travelers have confirmed: *In the vast Library there are no two identical books.* From these two incontrovertible premises he deduced that the Library is total and that its shelves register all the possible combinations of the twenty-odd orthographical symbols (a number which, though extremely vast, is not infinite[5]): in other words, all that it is given to express, in all languages.
>
> (54)

Everything we could possibly want to know—every true statement about the universe—is provided for us in the Library, if we can find the right book:

> When it was proclaimed that the Library contained all books, the first impression was one of extravagant happiness. All men felt themselves to be the masters of an intact and secret treasure. There was

no personal or world problem whose eloquent solution did not exist in some hexagon.

(54–55)

Indeed, there is not *just one* such: for every book that consists of only true statements, there are many others that provide the same statements in a different language, or in one of the number (quasi-infinite, again) of possible codes, known and unknown. By extension, *every* book can be considered a coded version of the "one true" book, by some unimaginably complex coding system:

> In truth, the Library includes all verbal structures, all variations permitted by the twenty-five orthographical symbols, but not a single example of absolute nonsense .... These phrases, at first glance incoherent, can no doubt be justified in a cryptographical or allegorical manner; such a justification is verbal and, *ex hypothesi*, already figures in the Library. I cannot combine some characters *dhcmrlchtdj* which the divine Library has not foreseen and which in one of its secret tongues do not contain a terrible meaning. No one can articulate a syllable which is not filled with tenderness and fear, which is not, in one of these languages, the powerful name of a god.
>
> (57)

Thus, the universe presents us with a fatal dilemma. There are coded messages waiting to be read, but—as in *Death and the Compass*—we have no *a priori* way of distinguishing them from meaningless noise:

> For four centuries now men have exhausted the hexagons .... I have seen them in the performance of their function: they always arrive extremely tired from their journeys; they speak of a broken stairway which almost killed them; they talk with the librarian of galleries and stairs; sometimes they pick up the nearest volume and leaf through it, looking for infamous words. Obviously, no one expects to discover anything.
>
> (55)

To conclude, let's look at a more recent work of Latin American literature: *Turing's Delirium* (2006) by Bolivian author Edmundo Paz Soldán, which (by no coincidence at all) uses a quote from The *Library of Babel* (the one from p. 57, cited in the preceding paragraph) as one of his epigraphs (we will address another of his epigraphs at the end of the chapter). Paz Soldán is a member of the "McOndo" movement, whose name signifies a new direction for his generation, turning from the magical realism of Garcia Marquez and his contemporaries (*One Hundred Years of Solitude* is set in the fictional town of Macondo) to the contemporary, connected, capitalist world, by alluding to both a new wave (*onda* in Spanish) and, of course, McDonald's.

This book is set in a somewhat fictionalized Bolivia, where Miguel Sáenz is a long-serving employee of the "Black Chamber" of the state security organization, established by dictator (and newly "elected" president) Montenegro to apply cryptanalysis to the suppression of revolutionaries. Sáenz has been nicknamed Turing by his boss Albert, in recognition of his uncanny codebreaking abilities. His "delirium"—more accurately, perhaps, delusion or obsession—is a compulsion to view *everything* he encounters in the world as a coded message, and to try to decode it. Some examples include:

> A billboard has been hung on a molding wall at the back of a lot: *Built Ford Tough*. An anagram in the last word: *Ought*. An ominous sign: imprisoned within those five letters is the word *go*. Ever since you were a child, you have felt that the world speaks to you, always, everywhere. That sensation has intensified in the past few months, to the point that you cannot read a sign or a word without thinking of it as a code, as a secret writing that needs to be deciphered .... Most people think literally and assume that *Built Ford Tough* means *Built Ford Tough*. You suffer from the opposite and spend entire nights awake, mourning the loss of the literal.
>
> (48)

> One of the police officers has a metal pin in his lapel with a red-and-white shield on it. What does it mean? This is the question you always ask yourself, the inevitable search for the lair in which meaning is hidden. Because you assume that nothing you rest your eyes on is what it seems to be; everything is a symbol, a metaphor, or a code for something else. The nervous way the police officer gesticulates ... the way his leather belt has skipped one of the loops on his belt.
>
> (200)

> [Y]ou cross Bacon Street in your gold Corolla and immediately think of William David Friedman, the American cryptanalyst who was convinced that Shakespeare's work contained secret phrases and anagrams that referred to the true author, Francis Bacon .... *It's no coincidence that Bacon Street is on my way*, you think, and without noticing you nearly run a red light at an intersection four blocks from the El Dorado.
>
> (47)

As with the previous examples we've examined, in both literature and science, the inability to distinguish between message and noise is at best a useless distraction, and more often carries potential for harm.

Note particularly the last of these three passages, in which Turing fails to heed a *real* coded message—the red light encoding "STOP!"—while lost in contemplating the significance of what obviously *is* a coincidence.

Furthermore—also as in those previous examples—the author puts the reader in an analogous position, providing many tantalizing features that *might* carry coded messages. Albert, Sáenz's boss (who suffers from advanced dementia—literal delirium), recalls having told Sáenz:

> I told him that I found inspiration in literature ... How to say the most obvious things using the most obscure words ... How to hide meaning in a forest of phrases ... Literature is the code of all codes ... it's one way of looking at the world. Of confronting the world. Of becoming immersed in the daily battle. Trying to see what's hidden. Covered by a layer of reality ... Trying to reach the core ... I told him to read "The Gold-Bug."
>
> (135)

Sáenz (a not uncommon Spanish surname) would have an English pronunciation close to "science"—or maybe "signs"; is one (or both) of those intended? A more elaborate possibility lies in the structural organization of the book. The point of view of the narration changes from chapter to chapter among *seven* different main characters: Sáenz himself; Ruth, his wife; Flavia, their daughter (and a skilled hacker); Albert; Ramírez-Graham, a Bolivian-American who worked for the National Security Agency and was recruited to come back and take over the leadership of the Black Chamber; Cardona, a judge, who once collaborated with the regime but now believes it was responsible for the death of his cousin when they were young, and is seeking revenge; and Kandinsky, a revolutionary (and another hacker).

Tantalizingly, the pattern of point of view alternation, sketched out in Figure 6.5, looks almost but not quite regular. But the sequence in the original work (Paz Soldán 2003) is *not the same*: several chapters have been moved around in the translation. Apparently that was done on purpose, as the author acknowledges the "excellent suggestions" of his editor in the "corrected and improved" English version, which would seem to suggest there *is* some significance to the ordering—but I have no idea what it might be. Perhaps the pattern represents a reflexive message about itself, namely: don't try to interpret me? Or perhaps (unlike Powers, whose arrangement of chapters in *GBV* we saw to be *very* significant) Paz Soldán meant nothing by it at all?

Another pattern: Sáenz's narration is given in the second person, while all the others (except for Albert's, which is in stream-of-consciousness-like first person) are in standard omniscient author's third person. Is there a message implicit in *this* choice? I believe so: it has the effect of highlighting the equivalence between the reader and Sáenz noted earlier:

	1	2	3	4	5	6	7	8	9	10	11	12	13	14	15	16	17	18	19	20	21	22
Sáenz	■															■						
Flavia		■					■				■						■					
Albert			■					■		■									■			
Ramirez				■					■											■		
Cardona					■								■		■						■	
Kandinsky						■																■
Ruth												■		■				■				

	23	24	25	26	27	28	29	30	31	32	33	34	35	36	37	38	39	40	41	42	43	44	ep
Sáenz	■						■						■				■						
Flavia		■								■								■				■	
Albert				■					■							■							
Ramirez					■	■						■		■						■	■		
Cardona																							
Kandinsky								■			■				■								■
Ruth			■																■				

Figure 6.5 Sequential narrative points of view for the successive chapters of *Turing's Delirium*.

both are always faced with the conundrum of message or noise. When Sáenz dies (shot by Cardona for complicity in the cousin's death) near the end of the book, he reflects (*you* reflect, since it is written in second person):

> Now you understand that your destiny was to attempt to decipher the codes that would lead you to discover the Code ... you think you can detect the patient work of a higher being, someone who is beyond all the codes and can explain them .... Your last thought is that you have ceased to think, that in reality you never thought, you were always delirious ... thought is a form of delirium, it's just that some deliriums are less offensive than others.
>
> (285)

I suggest we can read this as alluding to the *metaphorical* death of the reader, in the sense that the end of a book represents the end of one's life *qua* reader of the book, and the end of the quest to discover its Code.

Lastly, I call attention to yet another aspect of coding: translation, and the impossibility of perfect correspondence between the original and translated versions. Since the "coded messages" that Sáenz strives to interpret cannot possibly work the same in both Spanish and English, his "readings" come out very differently in the two versions, as for example:

> A billboard has been hung on a molding wall at the back of a lot: *Built Ford Tough.* An anagram in the last word: *Ought.* An ominous sign: imprisoned within those five letters is the word *go.*
>
> (48)

> En una mohosa pared posterior han desplegado un anuncio publicitario. *Camiones Ford.* Un anagrama en la primera palabra: *Es camino.* Un signo ominoso: apresada entre esas ocho letras, la palabra *Cain.*
>
> (61)

> Otis, six passengers, 1000 pounds. You stare at the name. You spell it out: O-T-I-S. Backwards: S-I-T-O. It is a message striving to break free, and it is destined only for you. I-O-T-S. *I'm Obliged To Say.* Who's obliged to say what?
>
> (9)

> Otis, *seis personas, cuatrocientos ochenta kilos.* Te quedas mirando el nombre. Lo deletreas: O-T-I-S. A la inversa: S-I-T-O. Un mensaje pugna por salir, y está destinado sólo para ti. S-T-O-I. *Soy Tu Oscuro Individuo.* ¿Quién es tu oscuro individuo?
>
> (20)

Note how the "ominous" sign "Cain" in the original becomes "go" (which doesn't seem so ominous) in the translation. Furthermore, if the author intended "Cain" to carry any special significance for the narrative, we have no way of recovering that by "decoding" the translation.

Another aspect of this "translation question" is presented by the character of Ramírez-Graham, who as a native Anglophone tends to slip between languages (such behavior is known as code-switching!)—but does so differently in the two versions:

> In my opinion, one of the main challenges to national security is cybercrime. Sí, incluso en Bolivia, mark my words .... Both the government and private industry increasingly depend on computers. Los aeropuertos, los bancos, el sistema telefonico, you name it.
>
> (27–28)

> Pienso que uno de los principales desafios a la seguridad nacional sera el cibercrimen. Sí, incluso en Bolivia, mark my words .... Tanto el gobierno como las empresas privadas dependen cada vez más de las computadoras. Los aeropuertos, los bancos, el sistema telefónico, you name it.
>
> (40)

Of course, such issues arise in *any* translated work, but they take on particular significance here, as the reader is virtually *forced* to be aware of the decoding and re-encoding that translation entails, adding another layer to the theme of hidden messages *vs.* noise that is central to the book. Connections between translation and L&S will be explored further in the next chapter.

Besides the Borges quote on p. 102, Paz Soldán used a line from Neal Stephenson's *Snow Crash* as another epigraph: "All information looks like noise until you break the code." Inverting that provides a much better motto for his book, as well as for this chapter of *my* book: All noise looks like information, if you believe there is a code.

Notes

1 This chapter is based on talks presented at the 2007 (Portland, ME) and 2010 (Indianapolis) SLSA meetings.
2 This refers to an anecdote (which I have come across in several places, but which nonetheless may well be apocryphal) that Powers cites (269): von Neumann skipped over part of a derivation during a lecture, proclaiming it to be obvious; when a student protested he couldn't follow that, von Neumann left the room for an extended period of time, returned to announce (with no elaboration) that yes, it *was* indeed obvious, and went on with his lecture.

3 One team who wrote on *GBV* (Herman and Lernout 1998) succeeded in interpreting the triplet scheme, apparently without any help from the author, even going so far as to suggest some possible dedicatees—James Clerk Maxwell for JCM, for example. I have no idea whether they got them right.

4 The last of these differs from the other two, in that it is more about how to live one's life than an expression of how to find truth in the world. Also, one might suggest that to be fully self-consistent, Thoreau should have omitted the second "simplify," although perhaps he did not consider consistency particularly essential, like his contemporaries Emerson ("A foolish consistency is the hobgoblin of little minds") and Whitman ("Do I contradict myself? Very well, then I contradict myself, I am large, I contain multitudes").

5 Naturally, the first time I read this story (inadvertently confirming my college English professor's suggestion that I treated literature too scientifically—see Chapter 1), I was compelled to figure out how big the Library would have to be to contain all those volumes. The calculation is straightforward: each book contains $410 \times 40 \times 80$ characters, or 1.3×10^6; each of these can have one of 25 identities, so the total number of books is 25 raised to the power of 1.3×10^6, which is approximately 10 to the power of 1.8×10^6. Each chamber contains 20×35 or 700 books, so the number of chambers is 1/700 of that number, which is essentially the same—1 followed by around a million zeroes. How does that compare with the known universe (ours)? The visible universe is a sphere whose diameter is around 93 billion light years; converting that to feet gives a total volume of about 10^{82} cubic feet. Information elsewhere in the story allows the volume of each chamber to be estimated as 400 cubic feet, so our known universe would hold around 10^{80} chambers: 1 followed by 80 zeroes, a number completely insignificant compared with Borges's Library.

7 Found in Translation[1]

My initial inspiration—provocation, actually—for this case study was a review article in the scientific literature, which argued for a particular approach to the visual representation of certain classes of molecules. The abstract ended with the following claim:

> This bonding description also provides a simple means to rationalize the theoretical predictions of the absence of M–M bonds in molecules such as $Fe_2(CO)_9$ and $[CpFe(CO)_2]_2$, which are widely misrepresented in textbooks as possessing M–M bonds.
>
> (Green *et al.* 2012, 11481)

This probably does not immediately strike one as promising material from which to launch an excursion into literature and science, and indeed my first reaction—one of strong disagreement—was purely on scientific grounds. However, as I framed a response to the authors (all friends and former colleagues), possible connections began to resonate. In particular, it started me thinking about *translation* as relevant not only to this specific dispute but to many aspects of scientific activity in general, and even more broadly as a useful metaphor for my own activity in literature and science.

In the latter regard, it may be noted that the concepts of translation and metaphor are closely related: the etymological derivation of the two words, from Latin and Greek, respectively, are essentially the same (Polizzotti 2018, 19).[2] The Latin word *translatio* comes from *trans* (= across) plus *latus*, the past participle of *ferre* (= to carry or to bring); the Greek word *metaphero* (μεταφερω) comes from μετα (*meta* = across) plus φερω (*phero* = to carry or to bear). Thus, both could be literally translated into English as "to carry across." Indeed, according to the online OED, one (obsolete) definition of "translation" is "transference of meaning; metaphor."

Douglas Hofstadter has proposed that translation is, at its core, equivalent to analogy—and as such can serve as a powerful learning enhancer. He perceives "translation as the faithful transport of some abstract pattern from one medium to another medium—in other words,

DOI: 10.4324/9781003197188-7

analogy" and goes on to say: "Analogy is really the name of the game. From a strong language you borrow skills and gain low-level competence in another one. From one puzzle you borrow techniques and solve others" (Hofstadter 1997, 45–47). Translation thus can serve as an illustration of the power of making connections between domains.

In this chapter I begin with a brief discussion of how translation manifests itself in science, both literally and metaphorically. I then give a basic introduction to chemical bonding and its representation, a necessary preface to the main topic, which is a detailed consideration of the specific scientific issue introduced earlier—how do we decide on a preferred representation of molecular structure?—and its close connection to problems in translation.

Science in Translation/Translation in Science

Translation in a literal sense—converting text from one language to another—has been crucial to the development of science over the centuries. Science claims to produce knowledge that is valid universally, not just locally, so clearly there cannot be insurmountable barriers to the transmission and understanding of that knowledge across borders. Equally clearly, the course of that development must be—or, at least, must have been—impacted by translation, which is never entirely unproblematic. It could be—and has been—argued that scientific knowledge in fact *can* be conveyed in a universal language, that of mathematics; but such an argument is not convincing. Not only is a large proportion of science irreducible to mathematics, but also many of the intended recipients are not able to fully understand or appreciate a purely mathematical presentation. More on this later in the chapter.

Translation has thus played a very important role in the history of science, which I do not have time or space to cover here. Fortunately, for those who are interested, two full-length (and first-rate) books have treated this issue in considerable depth: Scott Montgomery's *Science in Translation* (2000) and Michael Gordin's *Scientific Babel* (2015). As Montgomery poses the question: "[H]ow is knowledge rendered mobile? What makes it able to cross boundaries of time, place, and language?" (3). His treatment includes areas such as translation between Greek, Arabic, and Latin in pre-modern Western astronomy, and between Japanese and other languages in the evolution of modern Japanese science.

What about contemporary scientific practice? Would it be fair to say that translation has become significantly less central with the increasing dominance of English over the last half-century or so? That is the basic thesis of Gordin's book, and it certainly accords with my own experience. When I started my graduate work (in 1968), one strict requirement for the PhD was a working reading knowledge of two foreign languages (selected from German, French, and Russian), as demonstrated by the ability to translate (with a dictionary) a page from a scientific article.

Such a requirement made sense at the time; I would estimate that on the order of 20% of the articles I needed to read for my research were available only in German, and another 10% in French.

Russian was a different matter: although at least a comparable proportion of key articles had been written in Russian—over 20% in all fields of chemistry by 1970 (Gordin 2015, 217)—a majority of the ones I needed *were* available in English.[3] That was the consequence of an extremely large-scale program aimed at producing cover-to-cover translations of the most important Soviet journals, inspired by Cold War concerns that we absolutely *had* to keep track of their scientific developments, coupled with the perceived and/or real difficulties associated with Anglophones actually learning Russian (Gordin 2015, 251–266).

Today the situation is quite different: virtually any important scientific article is written in English, no matter what the authors' nationalities or mother tongues; and I would be surprised (I haven't checked) if any US universities still have a foreign language requirement for a graduate degree in science. The only time I have to resurrect my (badly faded) knowledge of French or German is when I am working on a problem in the history of chemistry. Gordin explores the reasons for, and advantages/disadvantages of, this takeover, but does not question that it has indeed become complete, while allowing for the possibility that it need not always be so: "The history of scientific languages ends here, until it no longer does" (Gordin 2015, 315).

While the very significant contributions translation has provided to science are now mainly of historical interest, science has been working on returning the favor, in the form of automated translation. The origin of that field can be traced back, again, to the Cold War–inspired need for translation of Russian scientific texts—particularly to a 1947 memorandum from Warren Weaver. A substantial program ensued; but even though efforts were limited to scientific/technical translation, thought to be the easiest target because of the universality of scientific language, results were at best marginal. As noted earlier, the need was ultimately satisfied in the old-fashioned way, by human translators (Gordin 2015, 213–266).

A well-known quote from the Weaver memo illustrates the misconception at the very core of the problem: "When I look at an article in Russian, I say: 'This is really written in English, but it has been coded in some strange symbols. I will now proceed to decode'" (Gordin 2015, 233). Could he really have believed that translation could possibly be so analogous to cryptanalysis? Perhaps he was only alluding to the use of a different alphabet for Russian; but it still reads as a remarkably unperceptive comment from someone farsighted in many other arenas.

More generally, while there has unquestionably been progress over the last 70 years, the tools I have seen are highly unsatisfactory—even in the limited domain of scientific texts. The last time I needed to translate a page or so of a German journal from around 1900 I began by running it through Google Translate, which gave me a reasonably helpful starting point; but I had to convert essentially *every* word and phrase into standard English using dictionaries and other resources. I doubt whether anyone foresees machines superseding humans as translators for anything other than extremely basic, highly restricted texts—certainly not for anything we would want to call "literature"—anytime in the near future.

Beyond these literal modes of participation in science, "translation" makes a number of appearances in more metaphoric form. The oldest of these is in physics—mechanics, specifically—where it refers to movement of a body from one position to another. Frequently the term is used in distinguishing classes of motion for an extended body. Movement of the center of mass of a body from one point in space to another is termed *translational* motion, in contrast to *rotational* motion of the body about its center of mass, and (for a non-rigid body) *vibrational* motion of parts of the body relative to others, while the center of mass remains stationary. In chemistry the latter two types of motion play a much more significant role than the former. Considering molecules as extended bodies (that is, made up of some number of atoms), vibrational and rotational motions can be detected by spectroscopy—the absorption of particular wavelengths of electromagnetic radiation in the infrared and microwave regions, respectively—which is used to provide important structural information.

Next, there is the "central dogma" of molecular biology, which proclaims that sequence information is passed from DNA to (messenger) RNA and thence to proteins. Those two steps are termed transcription and translation, respectively. It may seem that the latter is not really all that metaphoric but a rather literal usage, in the sense that the information contents of nucleic acids and proteins are effectively expressed with different "languages" using different alphabets. In DNA and RNA a "message" consists of a sequence of "words," each comprised of a triplet of "letters" chosen from four possibilities; each such three-letter word corresponds to a single amino acid in the protein sequence. Going from RNA to protein is thus a form of decoding, so it might seem natural to call it translation—even though, as noted earlier, any close association of decoding and translating is of questionable validity.

Of course, whenever we say anything about connections between genetics and information we are already being entirely metaphoric, starting from the very use of the term "genetic code." Indeed, our current usage of "translation" in this context took a little while to get firmly established. One early researcher introduced it rather hesitatingly: "mRNA chains are then copied (or, better, 'translated') into the amino-acid sequences of enzymatic proteins"; subsequent usages in the same article left out the quotation marks (Platt 1962, 168). Others reversed the descriptors, referring to generation of messenger RNA from DNA as translation (Ochoa 1962, 159) or protein synthesis as transcription (Beadle and Beadle 1966, 192). Clearly the language molecular biologists have *chosen* to use as they discuss their work among themselves, and present it to the world, is not "natural" at all, and surely has consequences. Lily Kay (among others) has discussed at length how these ubiquitous language/information metaphors influenced the development of the field over half a century (2000).

Lastly, phrases such as "translational medicine," "translational science," and "translational research" have come into vogue over the last 20 years or so. All seem generally to be taken to mean much the same thing: research that is intimately tied to potential clinical application, as opposed to being aimed mainly at enlarging our fundamental knowledge base (van der Laan and Boenink, 2015). Their earliest appearances date from around 1990, with usage growing steadily starting around the turn of the 21st century (as per the Google Books Ngram Viewer), although none had made it into the (online) Oxford English Dictionary as of the time of writing. Here are a couple of typical examples, one early and one more recent:

[T]he translation of such scientific knowledge into patient benefit is by no means clear cut in many instances, and the marriage between new discoveries in basic science and clinical practice—"translational medicine"—is a tremendous challenge.

(Geraghty, 1996)

Translational research means different things to different people, but it seems important to almost everyone .... For many, the term refers to the "bench-to-bedside" enterprise of harnessing knowledge from basic sciences to produce new drugs, devices, and treatment options for patients .... For others ... [it] refers to translating research into practice; i.e., ensuring that new treatments and research knowledge actually reach the patients or populations for whom they are intended and are implemented correctly.

(Woolf, 2008)

These terms thus have little to do with any literal sense of translation—indeed, their meanings seem to be quite nebulous and polyvalent—but their popularity is at least suggestive of a link between the concept of translation and scientific practice. I will now try to demonstrate a more explicit connection.

Models of Chemical Bonding

Our current system of visual representation of chemical structure basically developed during the second half of the 19th century, when organic chemists were engaged in chemically analyzing known compounds, synthesizing new ones, and deducing the rules governing how atoms combine to form molecules from their experimentally observed compositions.

Atoms of the various elements were found to be able to form a specific number of bonds to other atoms (in some cases more than one such number is possible): one for hydrogen, two for oxygen, four for carbon, and so on. These combining numbers are often referred to as valence. Molecules could thus be depicted using lines to show connectivity; before that, formulaic and structural representations were much more complex. Figure 7.1 shows some examples: hydrogen (known by then to consist of diatomic molecules), H_2; methane, CH_4; ethylene, C_2H_4; and benzene, C_6H_6. (From other work it was recognized that the four bonds to carbon in methane and related molecules do not all lie in a single plane, as indicated in the drawing; this is further discussed in the Appendix.) Although the evidence at that time for geometric arrangements of atoms into molecules was almost entirely inferential, later methodology—especially X-ray crystallography, first introduced by William H. and William L. Bragg (father and son) in 1912—provided much firmer empirical support.

Some substances exhibited compositions that did not immediately fit into this simple picture and required additional features, particularly the concept of multiple bonds. Thus, ethylene can satisfy the rule of tetravalent carbon—even though it appears there aren't enough atoms to provide four bonds to each carbon—if it is allowed that there may be a

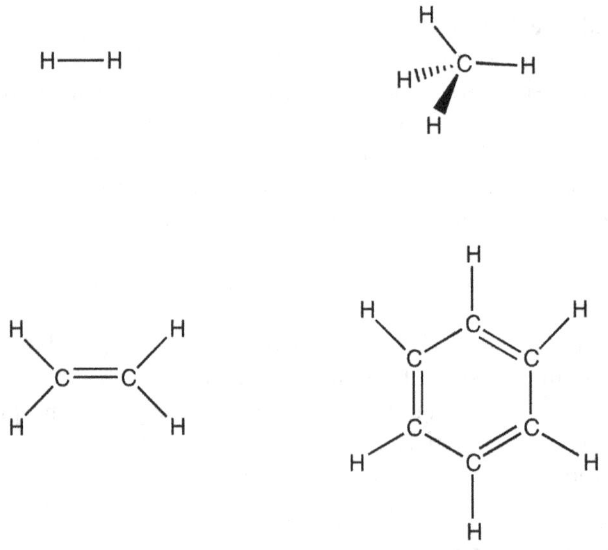

Figure 7.1 Simple structural representations of H_2, CH_4, C_2H_4, and C_6H_6, showing one bond to each H and four bonds to each C. The "wedged" lines for methane represent its non-planar structure: the four H atoms are tetrahedrally arranged about the central C, with the solid and dashed wedges conventionally indicating bonds above and below the page, respectively.

double bond between the two carbon atoms. Benzene presented a thorny problem, which was famously solved by August Kekulé in 1865 when (according to his subsequent account) he dreamed about a snake biting its own tail, and woke to formulate a ring structure, eventually elaborated to that shown in Figure 7.1.

What was not at all clear at the time, though, was what those lines actually *mean*—what the nature of the chemical bond between atoms might be. That understanding had to await several subsequent developments—Thomson's discovery of the electron (1897) and Bohr's orbital model of the atom (1913) in particular—which ultimately led to two main approaches to the description and portrayal of chemical bonding and structure in molecules that persist to this day.

The first, called valence bond (VB), basically treats each bond between two atoms as corresponding to a pair of electrons shared between them, represented in much the same way as the images in Figure 7.1 (with additional features for more complex molecules). These are usually called Lewis structures, after G. N. Lewis, who introduced them in the early 20th century. This sharing is energetically favorable, thus holding the molecule together. Electrons that can participate in bonding are called valence electrons—hence the name. This VB approach is probably the more familiar to most people; it was developed and popularized most notably by Linus Pauling in his seminal 1939 book *The Nature of the Chemical Bond*. VB is most often used qualitatively, although it can involve mathematical treatment.

An alternative model was developed beginning in the late 1920s by Robert Mulliken and other physical chemists, called the molecular orbital (MO) approach. MO does not identify particular bonds with electron pairs. Instead, it treats the atomic orbitals (AOs) of Bohr's model on each atom as interacting with one another to form combinations that extend over the entire molecule, and the valence electrons of the atoms occupy the most energetically favored MOs, thus accounting for the nature and stability of the chemical bonding. Computational chemists have developed increasingly sophisticated techniques for carrying out such calculations, which in favorable cases can give highly accurate predictions of molecular structure, as well as other observables such as reactivity, spectroscopy, *etc*. These MOs are mathematical constructs just like the AOs, and the most "correct" way to present the computational results would be as a set of formulae called wave functions. But the non-specialist often finds it hard to interpret those in structural terms, so more often we use diagrams to depict the energy levels and spatial distribution visually. Figure 7.2 displays sample versions of MO diagrams for two of the simplest cases, H_2 and CH_4, whose VB representations were included in Figure 7.1.

We can already see, even for such a relatively simple molecule as methane, that while the MO approach offers greater informational content, it does not make it so easy to readily grasp structural features—a challenge that increases dramatically with the size and complexity of the

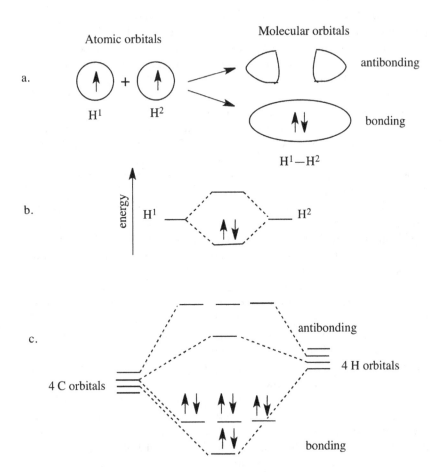

Figure 7.2 Molecular orbital representations of simple molecules. (a) Atomic orbitals of two H atoms, each providing one electron (shown as arrows), combine to form two molecular orbitals (MOs), called bonding and antibonding. (b) Energy diagram showing that the more stable (lower in energy) bonding orbital is occupied by the two electrons. (c) Energy diagram of the molecular orbitals of methane, CH_4; each of the four atomic orbitals of C (not shown) provides one electron, and these combine with those of the four H atoms (each providing one electron) to form eight MOs, four bonding (occupied by the eight electrons) and four antibonding (empty).

molecule. A common phrase deployed in the debate between the two alternatives (which was often vigorous during the first few decades after their introduction) was that VB is too good to be true (*i.e.*, easy to use but non-rigorous), whereas MO is too true to be good. We now recognize that each has advantages for particular usages, and most chemists in most situations are able to switch comfortably between them as appropriate. However, there are cases where such switching is accompanied by ambiguity and/or potential confusion—just as, I would like to suggest, translating from one language to another can be fraught with ambiguity. I will develop that analogy in the remainder of this chapter.

Translating Molecular Structure

Let us now return to the excerpt from a scientific paper that opened the chapter, which I recopy here:

> This bonding description also provides a simple means to rationalize the theoretical predictions of the absence of M–M bonds in molecules such as $Fe_2(CO)_9$ and $[CpFe(CO)_2]_2$, which are widely misrepresented in textbooks as possessing M–M bonds.
>
> (Green *et al.* 2012)

The molecule at issue, $[CpFe(CO)_2]_2$, comes from a class of compounds known as metal carbonyls. The first example, $Ni(CO)_4$, was discovered around 1890; subsequently an immense number (thousands, easily) of such compounds, containing one or more transition metal (the elements from the middle part of the periodic table) atoms bonded to one or more carbon monoxide molecules, with or without additional groups, have been prepared. There is a strong affinity between CO and transition metals; a familiar manifestation of that affinity may be seen in the toxicity of CO, which acts by binding to the iron (a transition metal) center in hemoglobin, thus blocking the transport of oxygen. A substantial fraction of known metal carbonyls contain more than one metal atom, which are held together by a direct electron-pair bond, as shown in the examples of Figure 7.3.

[CpFe(CO)$_2$]$_2$ was first prepared (at Harvard: my own PhD institution!) and structurally characterized in the 1950s. As the quote at the beginning of this chapter indicates, in the VB mode of representation it is almost always depicted as on the left side of Figure 7.4, in research and review articles as well as textbooks. Comparison with the examples of Figure 7.3 reveals two key points. First, in contrast to those, two CO

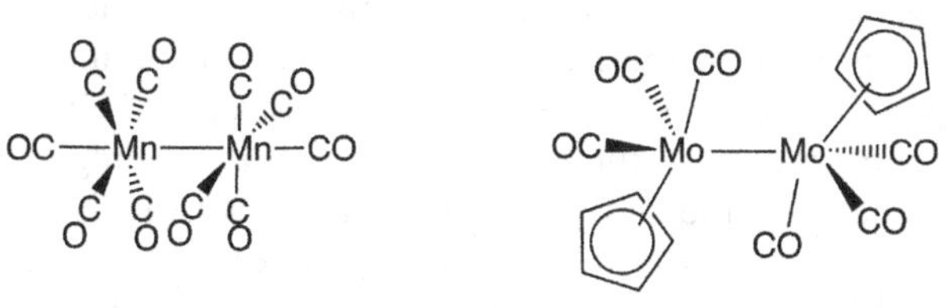

Figure 7.3 VB representations of two dimeric metal carbonyls, $Mn_2(CO)_{10}$ (left) and $[CpMo(CO)_3]_2$ (right). Cp is an abbreviation for cyclopentadienyl, the C_5H_5 group, which is conventionally represented in a VB diagram by a pentagon with a circle inscribed therein.

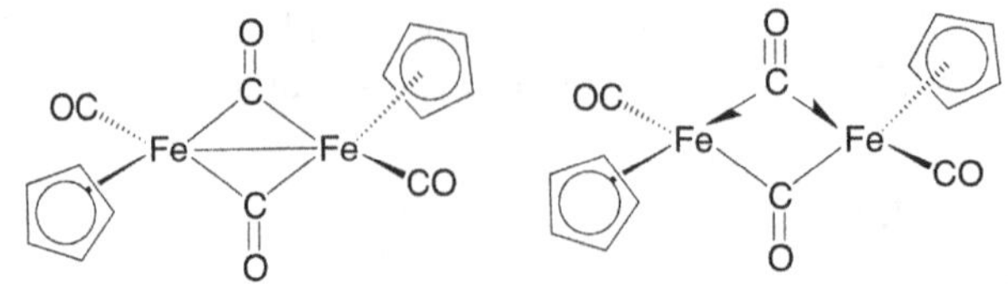

Figure 7.4 Two alternate VB representations of the [CpFe(CO)₂]₂ molecule: the traditional version (left) and that preferred by Green *et al.* (right).

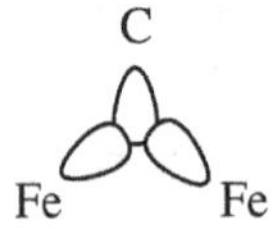

Figure 7.5 A molecular orbital generated by combining atomic orbitals from 2 Fe and 1 C atoms, termed a 3-center bond.

groups are bonded to *both* metal atoms: they are said to be bridging carbonyls. Second, as in the other examples, there is an explicit metal-metal (M-M) bond drawn between the two Fe centers.

In contrast, Green *et al.* reject the standard representation in favor of that shown on the right side of Figure 7.4. The latter differs from the former in several ways, two of which are of particular interest for this discussion. First, the two bonds between one of the bridging CO groups (the one on top) and the two Fe atoms are shown as "half-arrows" instead of simple lines; and second, *no* bond is drawn between the two Fe atoms: Green *et al.* call the inclusion of an M-M bond a misrepresentation. I will now try to show that both of these features evoke issues that arise in translating between languages.

First: what does the half-arrow signify in the VB approach? Well, traditionally, nothing at all. It was introduced by Green *et al.* to represent a not uncommon bonding situation in complex molecules: a so-called three-center two-electron bond. This is much easier to explain in MO language: it is a combination of three atomic orbitals, one from the C atom of CO and one from each of the Fe atoms, to form a single molecular orbital, which is occupied by two electrons, as sketched in Figure 7.5. (This may be contrasted with the description of the bonding of the bridging COs in the left-hand drawing, which as shown involves two-center two-electron bonds (like that in H_2) to connect each C to each Fe.) But standard VB "language" did not include any conventional means of depicting such an arrangement, so Green *et al.* basically invented one. We thus have a situation that is straightforwardly accounted for in one model (MO), while the other (VB) requires introducing a new, foreign symbolic representation.

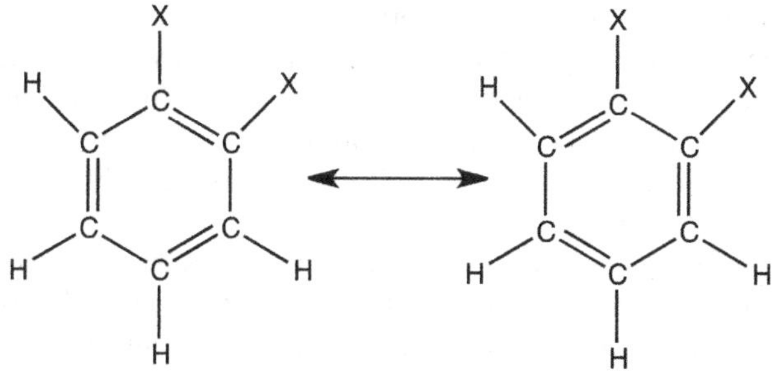

Figure 7.6 The two forms of a disubstituted benzene molecule in the VB representation.

This sort of imperfect correspondence between VB and MO descriptions is fairly common. A much older—and more familiar—example arises from the benzene molecule discussed earlier in this chapter. It is easy to see that the VB picture in Figure 7.1 implies that the bonds between adjacent carbon atoms are not all the same—there are alternating single and double bonds, so the structure should not be a *regular* hexagon. But in fact it is! The definitive proof of that had to await the development of crystallography and other methods of structural characterization in the 20th century; but even in the 19th century chemists understood that the benzene structure in Figure 7.1 was not an adequate representation. How did they know? If we make a new chemical species by replacing two H atoms of C_6H_6 with some other substituent X, *two* distinct compounds with adjacent X atoms should be obtainable, as shown in Figure 7.6, and that was already known *not* to be the case.[4]

In the MO model this presents no difficulties: the MOs generated by combining six AOs from six C atoms result in an overall bonding scheme that is completely symmetric with respect to the hexagonal ring, so all the C-C bonds are equivalent. But how could we express this in VB? Linus Pauling solved it by introducing a concept which he called resonance, represented by the double-headed arrow shown in Figure 7.6. It is important to understand that this representation does *not* mean that the two forms exist separately and rapidly interconvert (that situation is physically conceivable, but is not the case here); rather, that the actual structure is neither, but instead a hybrid or average of the two. Students (even some professionals, occasionally) are often careless and/ or confused about that distinction, which is arguably a consequence of using a somewhat artificial feature that does not fit so comfortably into the basic content of the VB approach—and, also, arguably partly due to Pauling's choice to call it "resonance." As we have seen earlier, that term typically refers to a phenomenon involving some form of oscillation, which could induce a subconscious (or even conscious) belief that the molecule does oscillate between the two structures drawn. (British physical organic chemists preferred "mesomerism" to describe the situation, but that never caught on, and could well have been at least as problematic, seeming to imply independent isomers.) This exemplifies a pitfall warned against in Chapter 4—that drawing inapt metaphoric or analogical connections between concepts from different realms can confuse instead of clarify—as well as another illustration of how difficult it is to separate science from language.[5]

Similarly imperfect correspondences are much more common—ubiquitous, in fact—in the realm of translation. We invariably find aspects of one language that cannot be made to correspond straightforwardly to those of another. Common examples include verb tenses, gendering of pronouns, *etc.* There are usually *reasonably* comfortable options for dealing with the problems thus encountered, but sometimes translators decide they need to "go outside the box" by making use of—or even creating—a non-standard device. I still recall my first encounter (in high school) with the following passage:

> "*Buena suerte*, Don Roberto," Fernando said ....
> "*Buena suerte* thyself, Fernando," Robert Jordan said.
> "In everything thou doest," Agustín said.
> "Thank you, Don Roberto," Fernando said, undisturbed by Agustín.
> "That one is a phenomenon, *Inglés*," Agustín whispered.
> "I believe thee," Robert Jordan said. "Can I help thee? Thou art loaded like a horse."
>
> (Hemingway 1940, 407)

My immediate reaction was: what on earth was Hemingway doing here? I finally understood that he (a) wanted to make it clear that the conversation was in Spanish (which should have been clear enough, given his inclusion of actual Spanish words!) and (b) felt compelled to deal with the fact that contemporary English does not differentiate between familiar and formal in the second person, unlike Spanish (and other languages commonly translated into English). So he chose to express the use of the familiar in his characters' spoken Spanish by means of the virtually completely archaic English forms "thou" and "thee." A poor decision, I thought then—and still do: it sounds so awkward and unusual as to more than obviate any potential benefit.

An example of *creating* a special device to handle the same problem is found in a recent translation of Françoise Sagan's *La Chamade*. Here's an excerpt:

> "What really gets me, personally, is the 'honest' style: 'Please forgive me, I thought I loved you, but I was wrong. It's my duty to tell you.'"
> "Surely that hasn't happened to You very often," said Antoine.
> "Thanks a *lot!*"

"What I meant is, You surely haven't given many men the chance to say such a thing to You. Your bags would have already been in the taxi!"

(Hofstadter 2009a, 19)

As Hofstadter explains in a preface (Hofstadter 2009b, 90–92), a French author's use of the familiar *tu vs.* the formal *vous* can carry important implications about relationships—informational and emotional content that Hofstadter was loath to give up, but which is pretty much untranslatable if one is restricted to standard English. Hence, he opted to use lower- and upper-case letters—"you" and "You"—to indicate *tu* and *vous*, respectively. Is this trick effective? Is it legitimate? That's open for debate—I'm guessing that few, if any, translators will follow his lead—but Hofstadter has certainly thought deeply about translation, and is entitled to try to make his case for it. In any case, though, I suggest that what he has done here, like Hemingway's thou- and thee-ing, is quite analogous to what Green *et al.* and Pauling did in inventing new symbolic representations for special bonding situations.

Let us now turn to the other issue with the VB picture: the presence/absence of an explicit M-M bond. Why do Green *et al.* omit it? Quantitative computational versions of the MO approach calculate the location of electrons within a molecule (or, more precisely, the spatial distribution of electron density, since Heisenberg taught us long ago that quantum mechanical particles cannot be precisely located). Most such calculations show no significant electron density along the Fe-Fe axis, and therefore Green *et al.* proclaim that any VB picture that *does* show a line there is a misrepresentation; instead, they draw it as shown on the right side of Figure 7.4 (Green *et al.* 2012).

My initial objection to their position was strictly from a chemist's point of view: that it, in turn, misrepresents (or fails to represent at all) features of the molecule that are arguably at least as important as the precise localization of electrons. In particular, it fails to capture a key point about chemical reactivity—which, after all, is the main thing that chemistry is all about! When we draw a line between two atoms to represent a two-electron bond, we imply (among other things) that at least in principle the bond could be split symmetrically, with one of those two electrons remaining with each half, generating two odd-electron fragments. And in fact, as shown in Figure 7.7, under irradiation with light (signified by the hv over the arrows), $[CpFe(CO)_2]_2$ *does* undergo such a fragmentation, just as do other dimeric metal carbonyls. The latter clearly have a metal-metal bond: there is nothing else to hold them together! All three of these compounds (and many others likewise) behave entirely analogously. By including the bond in the representation of $[CpFe(CO)_2]_2$ we foreground that pattern of reactivity; the alternate representation, without the bond, conceals it.

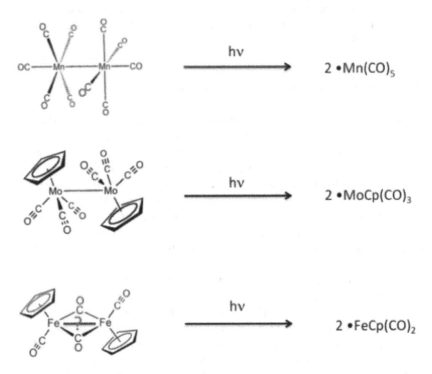

Figure 7.7 Cleavage of several dimeric metal carbonyl complexes into two equal fragments under photoirradiation.

Arguments can thus be made for either of the alternative VB representations in Figure 7.4. How can we decide which is better? The answer must depend upon which aspect(s) of the molecular description we most care about. If distribution of electron density is of prime import- ance, then the no-bond version might be considered a more accurate depiction. Of course, as noted earlier, we only know that distribution by means of an MO calculation, so an MO representation would seem to be much more appropriate in that case. Several different forms of such representations are shown in Figure 7.8. It should be apparent that while they may well convey substantially more information than (either of) the line drawings of Figure 7.4, they are not nearly so readily amenable to structural visualization. Those of us who are not completely comfortable with the highly mathematical representations may well prefer to convert them to a more easily grasped pictorial form—*i.e.*, a VB picture. As we have (at least) two different ways to do that, asking which representa- tion is "better" is equivalent to the question of how best to effect that conversion.

It was about at this point in my considerations that I recognized another connection between this mini-debate in chemistry and

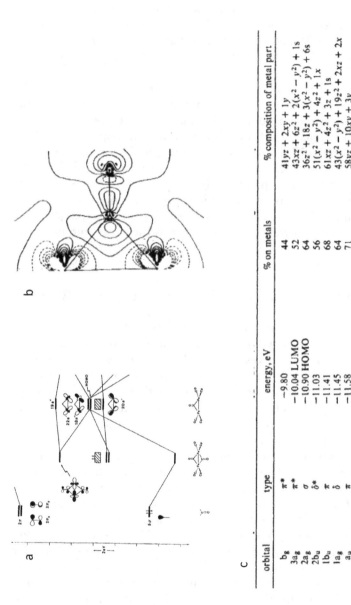

Figure 7.8 Three different MO representations of the [CpFe(CO)]₂]₂ molecule.

Adapted from (a) Bursten and Cayton 1986; (b) Bénard 1979; (c) Jemmis *et al.* 1980. Copyright American Chemical Society.

orbital	type	energy, eV	% on metals	% composition of metal part
b_g	π^*	-9.80	44	$41yz + 2xy + 1y$
$3a_g$	π^*	-10.04 LUMO	52	$43xz + 6z^2 + 2(x^2 - y^2) + 1s$
$2a_g$	σ	-10.90 HOMO	64	$36z^2 + 18z + 3(x^2 - y^2) + 6s$
$2b_u$	δ^*	-11.03	56	$51(x^2 - y^2) + 4z^2 + 1x$
$1b_u$	π	-11.41	68	$61xz + 4z^2 + 3z + 1s$
$1a_g$	δ	-11.45	64	$43(x^2 - y^2) + 19z^2 + 2xz + 2x$
a_u	π	-11.58	71	$58yz + 10xy + 3y$

translation in literature. It began with my recollection of Galileo's famous proclamation: "The book of nature is written in the language of mathematics."[6] MO representations are much more mathematical than the VB picture; but we may not be so fluent in that language as to be completely comfortable reading directly therein. Even such a major scientific figure as Michael Faraday felt that way (note that he uses the word "translating"!):

> There is one thing I would be glad to ask you. When a mathematician engaged in investigating physical actions and results has arrived at his own conclusions, may they not be expressed in common language as fully, clearly, and definitely as in mathematical formulae? If so would it not be a great boon to such as I to express them so?—translating them out of their hieroglyphics, that we also might work upon them by experiment .... If this were possible would it not be a good thing if mathematicians, writing on these subjects, were to give us their results in this popular, useful, working state, as well as in that which is their own and proper to them.
>
> (Faraday [1857] 2008, 305–306)

as did, apparently, several early 20th-century physicists (Whitworth 2001, 31).

In much the same sense, we can think about VB and MO as two different languages that we use to talk about molecular structure, and recognize that it is often necessary to "translate" from one to the other. This struck me as a most appropriate analogy, because the process of converting a mathematical formulation to one that is less so, as in going from MO to VB, raises the same sorts of issues that are common in literary translation. There are always priorities to be decided, and consequently choices to be made.

That recognition, in turn, reminded me that some years previously I had attended a panel on literary translation, at which one of the panelists—a translator from Italian—discussed several recent translations of Dante's *Divine Comedy*. During the ensuing Q&A I brought up a much older version by John Ciardi, which was the one I knew from my college days (the most commonly used at that time, I believe), and asked what he thought of it. I was rather taken aback by his response. I thought he might say it wasn't to his taste, or that he considered it inferior to the ones he preferred, but no: he proclaimed it *completely unacceptable*! Such a statement recalls that of Green *et al.* pronouncing the common VB representation of $[CpFe(CO)_2]_2$ to be a misrepresentation, not just a poorer choice.

What *was* that panelist's objection to Ciardi? The fatal deficiency, according to him, is already apparent in the first couple of stanzas of the original Italian and the translation in question:

Nel mezzo del cammin di nostra vita
mi ritrovai per una selva oscura,
ché la diritta via era smarrita.

Ahi quanto a dir qual era è cosa dura
esta selva selvaggia e aspra e forte
che nel pensier rinova la paura!
(Dante Alighieri 1317)

Midway in our life's journey, I went astray
from the straight road and woke to find myself
alone in a dark wood. How shall I say

What wood that was! I never saw so drear
so rank, so arduous a wilderness!
Its very memory gives a shape to fear.
(Ciardi 1954, 28)

The problem is Ciardi's decision not to follow the *exact* rhyme scheme Dante devised for the Comedy, called *terza rima*: the first and third line of each stanza rhyme with each other and the second line of the previous stanza (ABA BCB CDC ....). As can be seen, Ciardi doesn't *quite* manage that. In his translation the first and third lines of each stanza do rhyme with each other, but not with the second line of the previous stanza (ABA CDC EFE ....). For the panelist, that choice was enough to invalidate the whole translation.

There are of course *many* criteria one might use to assess a translation—especially of verse. They range from obvious ones, such as faithful representation of meaning, rhyme, meter, *etc.*, down to much more subtle aspects, such as keeping content correlated with position. (Ciardi doesn't do that either: the first half of the first line of the second stanza in the original is transposed to the first stanza in the translation. Should we care?) To disqualify it on the grounds of just one—any one—is to make an extremely strong value judgment about the relative importance of those criteria.

Furthermore, considering how impoverished English is in rhyming opportunities compared with Italian, it's far from clear to me that rhyming *should* be the number one criterion. According to Wikipedia (accessed in 2014) there have been many more than 100 translations of the *Commedia* into English over the years; only about a quarter of them even try to employ *terza rima*. Here are a couple of examples. The first (a recent one) is in strict *terza rima* but completely abandons Dante's regular meter; the second (a 19th-century version by Longfellow) is entirely unrhymed, but fairly faithful to the rhythm:

In the middle of our life's way
I found myself in a wood so dark
That I couldn't tell where the straight path lay.

Oh how hard a thing it is to embark
Upon the story of that savage wood,
For the memory shudders me with fear so stark
                              (Zimmerman 2003)

Midway upon the journey of our life
I found myself within a forest dark,
For the straightforward pathway had been lost.

Ah me! how hard a thing it is to say
What was this forest savage, rough, and stern,
Which in the very thought renews the fear.
                              (Longfellow 1865)

Which should we prefer? I happen to like the second—it sounds much better to me—but I certainly wouldn't pretend to have an argument that should convince everyone to agree with me.

Clearly, issues that arise in literary translation and in deciding how best to portray a molecular structure are closely akin to each other. The problematics of translation between languages apply to science just as they do to literature. How shall I "translate" the mathematical representation of my molecule into a pictorial representation? It depends on what aspects I am most concerned with portraying, as in Figure 7.4, where the with-bond version on the left tells us about reactivity, while the no-bond version may more accurately locate electron density. Different people can quite legitimately have different preferences, as they can for translations. What *is* to be avoided, in my opinion, is dogmatism: one should *not* proclaim one drawing or the other a misrepresentation, just as one should not proclaim a translation unacceptable, because of a choice that doesn't happen to agree with one's needs or preferences.

I suggest that keeping these connections between literature and science somewhere near the forefront of one's mind can have a significant and beneficial impact on scientific practice, even in such a "hard" scientific area as physical chemistry. There are potential advantages in pedagogy, as mentioned earlier: students often find it difficult to deal with concepts from different explanatory models that do not correspond readily to one another. Perhaps pointing out the parallels between such a situation and translation between language could help reduce the confusion.

It could also serve as a valuable reminder that there is rarely a single, incontestable choice of how best to describe a scientific fact or concept. In this regard, we might take note of Nabokov's argument in the preface to his translation of *Eugene Onegin*, that a proper translation should subjugate *both* rhyme and rhythm as priorities in favor of a hyper-literal rendering of the meaning:

In transposing *Eugene Onegin* from Pushkin's Russian into my English I have sacrificed to completeness of meaning every formal element save the iambic rhythm ... in the few cases in which the iambic measure demanded a pinching or padding of sense, without a qualm I immolated rhythm to reason. In fact, to my ideal of literalism I sacrificed everything (elegance, euphony, clarity, good taste, modern usage, and even grammar) that the dainty mimic prizes higher than truth.

(Nabokov 1964, *x*)

His dogmatic stance, I believe, has received little support from translators and readers alike.[7] One might think it is more compatible with a scientist's viewpoint—that accurately conveying "truth" must always supersede stylistic issues—but even in the realm of science such a commitment to straightforward expression of meaning is not sustainable. As two thoughtful commentators have expressed it:

[T]here is no single correct analysis of the complex entities of chemistry expressed in a single adequate language, as various reductionist scripts require; and yet the multiplicity and multivocality of the sciences ... do not preclude but in many ways enhance their reasonableness and success ... We understand the reality whose independence we honor as requiring scientific methods which are not univocal and reductionist precisely because reality is multifarious, surprising, and infinitely rich.

(Grosholz and Hoffmann 2012, 223)

Surely, in such a reality, literature and science should not be separate pursuits, but rather, should exercise their great potential for being mutually supportive.

Notes

1 This chapter is based on a talk presented at the 2016 (Atlanta) SLSA meeting. A (considerably) more technical version may be found in Labinger (2014).

2 Since structural chemistry will be a main subject of the second half of this chapter, it may also be worth noting that the two prefixes *trans* and *meta* are both used therein to designate spatial relationships, indicating substituents that are in some sense "across" from one another, as shown in Figure 7.9.

Figure 7.9 Left: a *trans*-disubstituted olefin; right: a *meta*-disubstituted benzene.

3 Nonetheless, I and my roommate, also a chemistry PhD candidate, did begin auditing a class in elementary Russian during my first semester at Harvard. After one week it became clear that (a) the instructor (himself a graduate student) absolutely required that we master *cursive* written Russian—totally unnecessary for our purposes—before going on to grammar and vocabulary; and (b) neither of us could tell one cursive Russian letter from another without extreme exertion. So that was the end of that attempt.

4 There *are* several compounds of formula $C_6H_4X_2$, known as isomers, but that is a consequence of different arrangements of the X atoms with respect to one another, not with respect to nonequivalent single and double bonds. We saw the *meta* isomer in Figure 7.9; the other two are called *ortho* and *para*, with the Xs adjacent and directly opposite, respectively.

5 Interestingly, Pauling's resonance concept has had some extra-scientific repercussions—not in the literary realm, but in politics! It was attacked in the Soviet Academy of Sciences in the 1950s:

> The Lysenko-era Russian researchers ... had for two years been tearing away at Pauling's "reactionary, bourgeois" chemical ideas, especially his use of idealized resonance structures with no real independent existence. Resonance theory, it was decided, was antimaterialistic and hence anti-Soviet ... "pseudo-scientific" and "idealistic" and should be rejected.
>
> (Hager 1995; for more details see Graham 1964)

That development was more than a little ironic, considering that Pauling was perceived in the US as leaning strongly to the left.

6 Of course, he *didn't* say that, not being an Anglophone; the actual quote is

> La filosofia è scritta in questo grandissimo libro che continuamente ci sta aperto innanzi agli occhi (io dico l'universo), ma non si può intendere, se prima non s'impara a intender la lingua, e conoscer i caratteri ne' quali è scritto. Egli è scritto in lingua matematica, e i caratteri son triangoli, cerchi ed altre figure geometriche, senza i quali mezzi è impossibile intenderne umanamente parola; senza questi è un aggirarsi vanamente per un oscuro laberinto.
>
> (Galileo 1623)

Translated into English:

> Philosophy is written in this grand book, the universe, which stands continually open to our gaze. But the book cannot be understood unless one first learns to comprehend the language and read the characters in which it is written. It is written in the language of mathematics, and its characters are triangles, circles, and other geometric figures without which it is humanly impossible to understand a single word of it; without these one is wandering in a dark labyrinth.
>
> (Drake 1960)

I think the familiar short version captures the meaning adequately.

7 In a commentary on the art of translation, Hofstadter takes Nabokov to task for such a "prosaic" approach (Hofstadter 1997, 257–267), but later proceeds to indulge himself in a comparably curmudgeonly diatribe about translations of Dante—though, to be fair, he does fully acknowledge that he is doing so, and explains why (528–549).

8 Entropy as Time's (Double-Headed) Arrow in Tom Stoppard's *Arcadia*[1]

What Is Entropy?

Almost as soon as the concept of entropy and the second law of thermodynamics were introduced, people began exploring the scope of their relevance. This is perhaps not surprising—if the second law tells us about things as small as the efficiency of a heat engine and as large as the ultimate fate of the universe, it seems reasonable to conclude that there are *no* matters that fall outside its domain. The breadth and depth of the implications of the second law have been emphasized by scientists, as in the following (somewhat tongue-in-cheek) comment:

> The law that entropy always increases—the second law of thermodynamics—holds, I think, the supreme position among the laws of Nature. If someone points out to you that your pet theory of the universe is in disagreement with Maxwell's equations—then so much the worse for Maxwell's equations. If it is found to be contradicted by observation—well, these experimentalists do bungle things sometimes. But if your theory is found to be against the second law of thermodynamics I can give you no hope; there is nothing for it but to collapse in deepest humiliation.
>
> (Eddington 1948, 74)

as well as by literary scholars. Eric Zencey borrows from philosopher Stephen Pepper the concept of "root metaphor"—a concept or fact that by analogy is used as the basis for understanding such a wide range of other areas that it can serve as the foundation of a comprehensive "world theory"—and argues that entropy can be considered as such a root metaphor (Zencey 1990).

Perhaps the most ambitious expression of that attitude was Jeremy Rifkin's (1980) book-length argument that *everything* in the world is determined by the second law—not just in a metaphoric sense, but rigorously:

The Entropy Law has a special power. It is so overwhelming that, once fully internalized, it transforms everyone it comes in contact with .... There will be those who will stubbornly refuse to accept the fact that the Entropy Law reigns supreme over all physical reality in the world. They will insist that the entropy process only applies in selective instances and that any attempt to apply it more broadly to society is to engage in the use of metaphor. Quite simply, they are wrong .... Every single physical activity that humankind engages in is totally subject to the iron-clad imperative expressed in the first and second laws of thermodynamics.

(6–7)

While a very large proportion of Rifkin's arguments for literal applicability are seriously overblown or just plain wrong—Zencey suggests the book might serve as evidence for a law of "intellectual entropy" (Zencey 1990, 192)—I will not examine them here. I cite him only as an illustration of the extremes to which some will go in using entropy to explain the world.

I present here just a brief account of how the concept of entropy arose and developed over the years (more detail may be found elsewhere, *e.g.*, in Müller 2007). The earliest statement of the underlying idea is often attributed to the French engineer Sadi Carnot, although some credit for recognizing the issue should go to his father Lazare, who discussed in an early 19th-century treatise the inherent tendency of mechanical engines to undergo loss of useful energy. The younger Carnot's analysis of steam engines noted that the work that can be obtained depends not just on absolute, but more importantly, on *relative* temperature—the difference between the temperature at which the steam is generated and that at which it is condensed back to water—and that heat can only move from a hotter to a colder body, unless a corresponding amount of work is consumed to push it in the other direction.

In the 1850s Rudolf Clausius provided concise statements of two laws of thermodynamics: first, that the energy of the universe is constant; second, that the *useful* energy of the universe is continually running down. To make the latter more precise he introduced the term entropy, from the Greek τροπη: a turning or transformation (a word that in the English form "trope" is ubiquitous in discussions of literature as well). Entropy, mathematically defined as the amount of heat transferred divided by the temperature ($\Delta S = Q/T$, where Δ signifies change, S entropy, and Q heat flow), could be shown to be a measure of energy that is *not* available for work, so that a formulation of the second law is that "the entropy of the universe tends to a maximum." On a much smaller scale, the statement also applies to an isolated system, but not to an open system that can exchange matter and/or heat with its surroundings. In the latter case the entropy of the system itself may decrease, but only at the expense of a larger increase elsewhere.

It was quickly recognized, by Lord Kelvin among others, that the two laws combined foretell the inevitable "heat death" of the universe: all energy must ultimately be distributed completely homogeneously, such that no useful work at all can be obtained. Entropy also came to be known as "time's arrow," as it implies that events can take place only in one direction—that in which entropy increases—in contrast to the mechanical universe of Newton's Laws, which appears to be perfectly reversible.

Somewhat later, in the 1870s, a statistical interpretation of entropy was developed by Ludwig Boltzmann, James Clerk Maxwell, and others, according to which entropy is a measure of the disorder or probability of the state of a system. Thus, for example, a container of gas in which all the molecules had been moved to one sector would be relatively more highly ordered, a much less probable situation than if they were uniformly spread throughout, and the former state would be of lower entropy. One quantitative form of this statement is Boltzmann's equation $S = -k_B \Sigma p_i (\log p_i)$, where k_B is a universal constant (which has come to be called Boltzmann's constant), and the p_i values give the probability of each of the system's accessible states.

While it may not be obvious that the thermodynamic and statistical versions of entropy are equivalent, there is a clear connection in that both point to a homogeneous distribution of matter and energy as the final, high-entropy destination. They *can* be shown to be rigorously equivalent for certain carefully defined ensembles of entities. But more generally, statistical entropy appears to have aspects that do not pertain to the earlier thermodynamic understanding, being probabilistic rather than causal. Also, it suggests some degree of subjectivity, since it is not always clear how to define relative order/disorder, and thus may be related to how much we understand about the system. American scientist J. Willard Gibbs commented:

> [T]he idea of dissipation of energy depends on the extent of our knowledge .... Dissipated energy is energy which we cannot lay hold of and direct at pleasure, such as the energy of the confused agitation which we call heat. Now confusion, like the correlative term order, is not a property of things in themselves, but only in relation to the mind which perceives them.
>
> (Denbigh and Denbigh 1985, 3)

This point was highlighted by a paradox introduced by Maxwell, who imagined a tiny being (which came to be known as Maxwell's Demon) sitting inside a closed system partitioned into two sections, with an opening that he could open or close at will to permit only fast-moving gas molecules to enter one side and slow ones the other. Since gas velocity is directly related to temperature, this would constitute moving heat from a colder to a hotter region without expending work, thus decreasing entropy in a closed system and violating the second law. Despite the fantastical nature of this concept, it occasioned an ongoing debate (that still persists, sometimes), with many explanations offered for why the second law was not violated.

In the mid-20th century—during the time of burgeoning interest in cybernetics and computation—another interpretation of entropy developed in the context of information theory. (The following is an extremely cursory summary; more thorough accounts may be found in Hayles 1990b or Denbigh and Denbigh 1985, 101–118.) Leon Brillouin suggested that the Demon had to manipulate information to carry out his sorting task, resulting in an increase in entropy that would more than compensate for the apparent decrease; he proposed "negentropy" as a new term for information. Around the same time, Claude Shannon, working at Bell Laboratories, tried to quantify the information content of a coded message (such as might be communicated electronically) in terms of probabilities: a message with no improbability, where every element can be confidently predicted, conveys no new information at all. He devised an equation for the information content, which came out as $S = - \Sigma \, p_i \, (\log p_i)$—the same form as Boltzmann's entropy formula. Hence, he decided to use the term entropy here as well, reportedly following physicist John von Neumann's suggestion: "It is already in use under that name ... and besides it will give you a great edge in debates because nobody really knows what entropy is anyway" (Denbigh and Denbigh 1985, 104). In Shannon's picture, contrary to Brillouin's, entropy and information have the *same* sign—more information = higher entropy. It should be emphasized, though, that both of these concepts involve formal definitions of information that need not have *anything* to do with meaning.

Does this idea of subjective entropy have any rigorous validity, as might be suggested by Shannon's mathematical relationships between "entropy" and information? At least one author—an astrophysicist—claimed in a *Scientific American* article that they *are* indeed equivalent and interchangeable:

> The thermodynamic entropy of Kelvin and Clausius and the statistical entropy of Boltzmann, Gibbs and Shannon have identical mathematical properties: they are aspects of a single concept. Entropy and information are related by a simple conservation law, which states that the sum of the information and the entropy is constant .... Thus a gain of information is always compensated for by an equal loss of entropy.
>
> (Layzer 1975, 60)

But those who have carefully looked into the relationship more typically conclude that it is at best an approximation, applicable only in certain highly restricted situations. More generally, its utility lies in the realm of metaphoric explanatory power:

Although information theory is more comprehensive than is statistical mechanics, this very comprehensiveness gives rise to objectionable consequences when it is applied in physics and chemistry .... It remains true, nevertheless, that information theory can be of value in an heuristic sense .... Notions about "loss of information" can sometimes be intuitively useful. But they can also, like the comparable concept of "disorder," give rise to mistakes .... It needs to be kept in mind that *thermodynamic* entropy is fully objective ... and the same must apply to any other "entropy" which is used as a surrogate.

(Denbigh and Denbigh 1985, 117)

More recent considerations of the second law have shifted the emphasis from destructive to *constructive* aspects of entropy. While not denying the long-term universal path to a completely disordered equilibrium state, Prigogine and Stengers (1984) focus upon the ability of what they call "dissipative systems" to *create* order. Systems close to equilibrium tend to undergo gradual, reversible change; but a system far from equilibrium has strong driving forces for irreversible change, and can increase its order by dissipating disorder (entropy) to its environment:

A new unity is emerging: irreversibility is a source of order at all levels. Irreversibility is the mechanism that brings order out of chaos. How could such a radical transformation of our views on nature occur in the relatively short time span of the past few decades? We believe that it shows the important role intellectual construction plays in our concept of reality.

(292)

Similarly, as a popular book on chaos theory put it, "Somehow, after all, as the universe ebbs toward its final equilibrium in the featureless heat bath of maximum entropy, it manages to create interesting structures" (Gleick 1987, 308). Schneider and Sagan (2005) provide a useful full-book-length exposition of this concept and its consequences.

Both thermodynamics and order out of chaos are major themes of Stoppard's *Arcadia*. Before examining how they feature in that play, though, let's look briefly at literary treatments of the second law more broadly.

Entropy in Literature

The question of how the second law applies to the social world and the human mind, and its interactions with the physical world, has received intense and enduring attention from the early days of the concept of entropy; and the preceding analysis does not deny that even approximate analogies may be of value, if only to provide mental stimulus. In this section I will highlight just a small selection of literary appropriations of the concept. An early example is Henry Adams's (1910) promulgation of the role of entropy in history:

> [T]he American professor who should begin his annual course by announcing to his class that their year's work would be devoted to showing in American history "a universal tendency to the dissipation of energy" and degradation of thought, which would soon end in making America "improper for the habitation of man as he is now constituted," might not fear the fate of Giordano Bruno, but would certainly expect that of Galileo.
>
> (85)

> The historian of human society has hitherto, as a habit, preferred to write or to lecture on a tacit assumption that humanity showed upward progress ... but this passive attitude cannot be held against the physicist who invades his territory and takes the teaching of history out of his hands.
>
> (87)

> As an energy [Man] has but one dominant function:—that of accelerating the operation of the second law of thermodynamics.
>
> (155)

Twentieth-century references include—somewhat unexpectedly—detective fiction, as in Lord Peter Wimsey's (appropriately whimsical) comment:

> What I like about your evidence, Miss Kohn, is that it adds the final touch of utter and impenetrable obscurity to the problem .... It reduces it to the complete quintescence of incomprehensible nonsense. Therefore, by the second law of thermo-dynamics, which lays down that we are hourly and momently progressing to a state of more and more randomness, we receive positive assurance that we are moving happily and securely in the right direction .... I have got to the point now at which the slightest glimmer of commonsense imported into this preposterous case would not merely disconcert me but cut me to the heart. I have seen unpleasant cases, difficult cases, complicated cases and even contradictory cases, but a case founded on stark unreason I have never met before.
>
> (Sayers 1932, 236)

and—less unexpectedly—science fiction. My first encounter with the term that I recall (although I'm sure I didn't understand at the time what it meant) was in a Heinlein short story, where two characters (actually the same character, meeting himself after a temporal displacement) argue about whether time travel is possible:

"Don't worry about it. The causation you have been accustomed to is valid enough in its own field but is simply a special case under the general case. Causation in a plenum need not be and is not limited by a man's perception of duration." Wilson thought about that for a moment. It sounded nice, but there was something slippery about it. "Just a second," he said. "How about entropy? You can't get around entropy." "Oh, for heaven's sake," protested Diktor, "shut up, will you? You remind me of the mathematician who proved that airplanes couldn't fly."

(Heinlein [1941] 1962, 75)

A more familiar sci-fi example is Isaac Asimov's short story "The Last Question," which *could* be taken to address the issue raised here: whether informational entropy—here represented in the form of a universe-sized computer—can compensate for, and reverse, the consequences of thermo-dynamic entropy, although it is far from clear that the author intended such an interpretation:

The last question was asked for the first time, half in jest, on May 21, 2061, at a time when humanity first stepped into the light .... How can the net amount of entropy of the universe be massively decreased? Multivac fell dead and silent. The slow flashing of lights ceased, the distant sounds of clicking relays ended. Then, just as the frightened technicians felt they could hold their breath no longer, there was a sudden springing to life of the teletype attached to that portion of Multivac. Five words were printed: INSUFFICIENT DATA FOR MEANINGFUL ANSWER.
. . . . . .

"Cosmic AC," said Man, "How may entropy be reversed?" The Cosmic AC said, "THERE IS AS YET INSUFFICIENT DATA FOR A MEANINGFUL ANSWER." Man said, "Collect additional data." The Cosmic AC said, "I WILL DO SO. I HAVE BEEN DOING SO FOR A HUNDRED BILLION YEARS. MY PREDECESSORS AND I HAVE BEEN ASKED THIS QUESTION MANY TIMES. ALL THE DATA I HAVE REMAINS INSUFFICIENT."
. . . . . .

Man's last mind ... looking over a space that included nothing but the dregs of one last dark star and nothing besides but incredibly thin matter ... said, "AC, is this the end? Can this chaos not be reversed into the Universe once more? Can that not be done?" AC said, "THERE IS AS YET INSUFFICIENT DATA FOR A MEANINGFUL ANSWER."

And it came to pass that AC learned how to reverse the direction of entropy. But there was now no man to whom AC might give

the answer of the last question. No matter ... And AC said, "LET
THERE BE LIGHT!" And there was light----

(Asimov [1956] 1969)

Among "mainstream" authors of fiction who have addressed entropy,
Thomas Pynchon is probably the most significant, starting with his story
of that title:

> [He] found in entropy or the measure of disorganization for a closed
> system an adequate metaphor to apply to certain phenomena in his
> own world. He saw, for example, the younger generation responding
> to Madison Avenue with the same spleen his own had once reserved
> for Wall Street: and in American "consumerism" discovered a similar
> tendency from the least to the most probable, from differentiation
> to sameness, from ordered individuality to a kind of chaos. He
> found himself, in short, restating Gibbs' prediction in social terms,
> and envisioned a heat-death for his culture in which ideas, like heat-
> energy, would no longer be transferred, since each point in it would
> ultimately have the same quantity of energy; and intellectual motion
> would, accordingly, cease.
>
> ([1960] 1984, 88–89)

It is ironic that Pynchon—or maybe just his protagonist?—recognizes the
importance of a "closed system" in his definition of entropy, but then
applies the concept to society, a system that is about as far from closed
as possible, raising the question of just how "adequate" is his chosen
metaphor. In the introduction to the collection in which the story was
reprinted, Pynchon expresses substantial regret for his early effort:

> Disagreeable as I find "Low-lands" now, it's nothing compared to
> my bleakness of heart when I have to look at "Entropy" .... Because
> the story has been anthologized a couple-three times, people think I
> know more about the subject of entropy than I really do .... But do
> not underestimate the shallowness of my understanding .... It is cold
> comfort to find out that Gibbs himself anticipated the problem, when
> he described entropy in its written form as "far-fetched ... obscure
> and difficult of comprehension."
>
> (1984, 12–14)

This apologia fails to acknowledge that in the intervening years he
had delved *much* further into the topic, including Shannon, Maxwell's
Demon, and more:

> He began then, bewilderingly, to talk about something called entropy
> .... But it was too technical for her. She did gather that there were two
> distinct kinds of this entropy. One having to do with heat-engines,

the other to do with communication. The equation, for one, back in the '30s, had looked very like the equation for the other. It was a coincidence. The two fields were entirely unconnected, except at one point: Maxwell's Demon. As the Demon sat and sorted his molecules into hot and cold, the system was said to lose entropy. But somehow the loss was offset by the information the Demon gained about what molecules were where …. "Entropy is a figure of speech, then," sighed Nefastis, "a metaphor. It connects the world of thermodynamics to the world of information flow. The Machine uses both. The Demon makes the metaphor not only verbally graceful, but also objectively true."

(Pynchon 1966, 72–73)

This passage seems rather equivocal on the ontological status of entropy: both a "verbally graceful" metaphor *and* "objectively true!" An interesting analysis of the science in *The Crying of Lot 49* by collaborating professors of English and physics (a laudable, and all too rare, enterprise) appeared a few years after the book (Lyons and Franklin 1973). Indeed, critics have made a small cottage industry out of tracing themes of entropy and chaos through Pynchon's later oeuvre, as well as that of his contemporaries (White 1991; Porush 1991). Bruni (2011) also gives a useful overview of thermodynamics in literature.

Science and the Theater: The Second Law in *Arcadia*

Plays with some scientific content or relevance go back at least to the early 17th century (Marlowe's *Doctor Faustus*, Jonson's *The Alchemist*), but they have become significantly more common in the last few decades. An excellent treatise on the subject both covers the history in general and examines a number of important examples in detail (Shepherd-Barr 2006); also, a recent compendium of more specialized essays (Shepherd-Barr 2020) has been published. I will not repeat much of that here, but I do want to emphasize how the use of this particular medium presents both opportunities and pitfalls in conveying scientific ideas. We saw (in Chapter 2) that Aldous Huxley was dubious about the compatibility of science and drama (he was more open to the idea of science in comedy); I ascribed that to his (mistaken) perception of science as "a matter of disinterested observation, unprejudiced insight and experimentation, patient ratiocination." But to the degree that transmitting scientific content requires some amount of explicit exposition (which will almost always be the case), the playwright must find a way of both achieving that and sustaining audience interest, preferably without descending into didacticism.

I say "the playwright," but of course a play is not *just* a text—a point Stoppard himself made at a talk I heard him give at Caltech, shortly after the first time I saw *Arcadia*:

[S]uppose you were to go into the campus bookshop and say, "I want *Pride and Prejudice*, please, and Beethoven's Fifth, and I'd like Warhol's Marilyn Monroe print, and I would like *Death of a Salesman* by Arthur Miller." [H]e gives you a bound stack of pages between two covers, and he gives you a circular disk, a flat thing, and then he gives you a kind of flat rectangular plane which goes on the wall. And then he gives you another stack of pages. And you say, "No, no. The Arthur Miller one is a *play*." And he'd say, "Well, yeah, that's how they come." There's something odd about this. I suppose a play is a text, but theater is an event.

(1994, 5–6)

Stoppard here is emphasizing the centrality of the entire theater experience—which depends upon the director, actors, production staff, *etc.*, along with the author—in communicating with the audience. Shepherd-Barr remarks: "There are many science plays whose scientific content is unassailable but whose theatricality is weak. They may teach science, but they do not make superb or even satisfying drama" (2006, 12). Of course, "theatricality" need not entail "grand spectacle." Perhaps the most successful science play of the last couple of decades, Michael Frayn's *Copenhagen*, employs only three actors on a nearly bare stage; but the science is so well integrated with the form and action that it virtually compels attention and involvement.

Tom Stoppard has been one of the leading figures of the English-language stage ever since his 1967 play *Rosencrantz and Guildenstern Are Dead*. Besides *Arcadia* (Stoppard 1993), two of his plays have strong science connections: the earlier (1988) *Hapgood*, which explores ambiguity by interweaving espionage and quantum physics; and the more recent (2015) *The Hard Problem*, which (in part) is concerned with explaining consciousness, the titular "hard problem."[2] *Arcadia* was first produced in London in 1993; I saw it there the following year (after it transferred from the National Theatre to the West End), and have attended three or four other productions subsequently. Clearly, not only I but many theater companies and audiences have found it very worthwhile.

The main scientific themes of the play are chaos, time, and entropy. To be sure, it is about *much* more than that: other themes featured prominently include English gardens, Romantic poets, straying wives, outraged husbands, obsessed academics, *etc*. But a significant component is Stoppard's image of how entropy and time work—at least in the world of the human mind—which is intriguingly different from more traditional views. One aspect that makes it particularly intriguing, as I will try to show, arises from the mode of presentation: not by overt exposition (although there is a good deal of explicit discussion of the scientific themes in the text) but, much more subtly, via the overall structure and staging of the play.

Arcadia is set in a single location—Sidley Park, an English country estate—but in two time periods; the action alternates between them, with Scenes 1, 3, and 6 set in 1809; 2, 4, and 5 today; and 7 in *both*—more on that later. There are three main plot components in the early period; most important is the ongoing interaction between Thomasina Coverly, the teenaged daughter of the manorial family, and her tutor, Septimus Hodge. The second centers around Ezra Chater, a (mediocre) poet who is a guest of the family, and whose wife (who is never seen) is remarkably promiscuous, having several sexual encounters with (at least) Septimus. Finally, the park surrounding the estate is undergoing a complete overhaul, from "the familiar pastoral refinement of an Englishman's garden" (whence, in part, the play's title) to something much wilder—"an eruption of gloomy forest and towering crag" (12)—under the direction of landscape architect Richard Noakes, and to the considerable dissatisfaction of Thomasina's mother, Lady Croom (at one point she dubs him "Culpability Noakes"). The other characters (Jellaby, the butler; Augustus, Thomasina's elder brother; and Captain Brice, Lady Croom's brother) have only minor roles with respect to the scientific theme.

In the contemporary period, there are three key characters, all academic types. Valentine Coverly, a scion of the same family, which still owns the estate, is a mathematician working on chaos theory—specifically population dynamics, as documented in the records of grouse hunting in the estate game books. Hannah Jarvis is a historian and houseguest, who is studying the history of the estate's gardens—with a particular interest in a mysterious hermit—at the request of the current Lady Croom (never seen, like Mrs. Chater). Bernard Nightingale, an English don, comes looking for support for a new theory about Byron, which he is about to announce at a meeting. There are also two younger Coverly children, Chloë and Gus.

The contrast between Classical and Romantic signaled by the relandscaping of the estate—which, according to Stoppard, was the first germ of his conception of the play (Kelly and Demastes 1994, 2)—is echoed, very early on in the play, in the scientific themes expressed by Thomasina, who is clearly *extremely* precocious. Within the first ten minutes we get this exchange:

THOMASINA: When you stir your rice pudding, Septimus, the spoonful of jam spreads itself round making red trails like the picture of a meteor in my astronomical atlas. But if you stir backward, the jam will not come together again. Indeed, the pudding does not notice and continues to turn pink just as before. Do you think this is odd?
SEPTIMUS: No.
THOMASINA: Well, I do. You cannot stir things apart.
SEPTIMUS: No more you can, time must needs run backward, and since it will not, we must stir our way onward mixing as we go, disorder out of disorder into disorder until pink is complete, unchanging and

unchangeable, and we are done with it for ever. This is known as free will or self-determination.

(4–5)

followed almost immediately by her asking Septimus whether she is "the first person to have thought of" this:

THOMASINA: If you could stop every atom in its position and direction, and if your mind could comprehend all the actions thus suspended, then if you were really, *really* good at algebra you could write the formula for all the future; and although nobody can be so clever as to do it, the formula must exist just as if one could.

(5)

The latter is essentially Laplace's comment about the ultra-ordered, Newtonian mechanistic universe (which Thomasina *would* have beaten by a few years, in 1809):

We ought then to consider the present state of the universe as the effect of its past and as the cause of that which is to follow. An intelligence that, at a given instant, could comprehend all the forces by which nature is animated and the respective situations of the beings that make it up, if moreover it were vast enough to submit these data to analysis, would encompass in the same formula the movements of the greatest bodies of the universe and those of the lightest atoms. For such an intellect nothing would be uncertain and the future, like the past, would be open to its eyes.

(Laplace [1814] 1998, 2)

while her "rice pudding" query anticipates the second law—particularly the statistical understanding thereof and "time's arrow"—and its undermining of classical Newtonianism.

Later in the play the implications of the second law are stated even more explicitly:

THOMASINA: [B]ad news from Paris! .... Newton's equations go forwards and backwards, they do not care which way. But the heat equation cares very much, it goes only one way. That is the reason Mr. Noakes's engine cannot give the power to drive Mr. Noakes's engine.
SEPTIMUS: Everybody knows that.
THOMASINA: Yes, Septimus, they know it about engines! (86–87)
SEPTIMUS: So, we are all doomed!
THOMASINA: (*Cheerfully*) Yes.
. . . . . .

SEPTIMUS: So the Improved Newtonian Universe must cease and grow cold. Dear me.

(93)

The "bad news from Paris" presumably alludes to the elder Carnot's treatise mentioned earlier. These exchanges represent the opposition between Laplace's eternal "ideal perpetual-motion machine" (Prigogine and Stengers 1984, 115) and the universe decaying towards its inevitable heat death—and if we go no deeper than these expository statements, then Stoppard seems to be accepting the latter, pessimistic view. But when we combine them with the interwoven contemporary story, we get a very different picture.

All three of the academics are engaged in projects aiming at reconstructing the past. Bernard has stumbled across a manuscript of Chater's found in Byron's library, which contains both an inscription from Chater to Septimus—hence the Sidley Park connection—and letters—with addressee unnamed—that he reads off to Hannah:

"Sir—we have a matter to settle. I wait on you in the gun room. E. Chater, Esq"

"My husband has sent to town for pistols. Deny what cannot be proven—for Charity's sake—I keep my room this day." Unsigned.

"Sidley Park, April 11th 1809. Sir—I call you a liar, a lecher, a slanderer in the press and a thief of my honour. I wait upon your arrangements for giving me satisfaction as a man and a poet. E. Chater, Esq"

(31)

In light of the fact that Byron left England for two years around that time (which *is* a known fact), and that there is no appearance of (the fictional) Chater in literary history after the manuscript mentioned, Bernard leaps to the conclusion that Byron was a guest at the estate in 1809, had an affair with Mrs. Chater, killed Chater in a subsequent duel and was forced to flee. He receives further support on learning (from Hannah) that Captain Brice married Chater's widow in 1810, and (from Valentine) that Byron was indeed a guest at Sidley Park at the crucial time.

The audience knows from early on that most of this is wrong—it was Septimus, not Byron, who had an affair with Mrs. Chater and was the recipient of the letters—but when Hannah and Valentine urge caution, Bernard refuses to accept it:

BERNARD: Is it conceivable that the letters were already in the book when Byron borrowed it?

. . . . . .

VALENTINE: Well, it's conceivable.

BERNARD: Is it *likely* that Hodge would have lent Byron the book without
  first removing the three private letters?

(56)

and proceeds to deliver his paper—only to be undone by Hannah's discovery in the garden book (which we learned about in an earlier 1809
scene) that in fact Chater sailed off to Martinique on a botanical expedition with Captain Brice and there died, bitten by a monkey. Hannah's
and Valentine's research projects are similarly confounded: Hannah is
unable to identify the hermit—it turns out to have been Septimus, who
abandons the world in despair after Thomasina is burned to death in
1812—and Valentine's mathematical project succumbs to excess noise in
the data.

What does this have to do with the second law? Since *Arcadia* presents
us with a single setting at two different times, we can examine how the
"system" evolves over time. There are several indications of evolution in the
forward direction being characterized by disorder and degeneration, most
notably the "improvements" of the Sidley Park gardens; also somewhat
veiled hints about the Coverly family going downhill: whereas Thomasina
is a child genius and Augustus is a budding young aristocrat, Chloë appears
to be pretty much an airhead, while Gus is mysteriously mute.

What happens in the reverse direction? A major focus of the play is on
the various attempts, especially Bernard's, to reconstruct the past. In the
informational world, metaphorically, these are *backwards* trips in time.
According to Prigogine and Stengers, as paraphrased by Hayles:

> When time goes forward there is a role for chance, because small or
> random fluctuations near a bifurcation point can cause a system to
> take a different path than it otherwise would …. But when time runs
> backward along the same track it took before, every juncture point is
> already predetermined, and hence chance can play no further part in
> the system's evolution.
>
> (Hayles 1990a, 98–99)

But in *Arcadia* the backwards time travelers are subject to exactly the
chance events and random fluctuations that Prigogine and Stengers deny!
Bernard in particular gets (nearly) everything wrong, because letters are
left in the wrong place, misleading inscriptions are misattributed, crucial
documents turn up at the wrong time, *etc.* So if we follow Brillouin in
equating randomness, disorder, and loss of information with increasing
entropy, we get the result shown in Figure 8.1: if entropy is "time's
arrow," in this play it points both ways!

How can a quantity increase in both directions? In the mental/per-
ceptual world it can: a familiar illustration is shown in Figure 8.2, where
one gets from point A to point B by going uphill, and from point B to

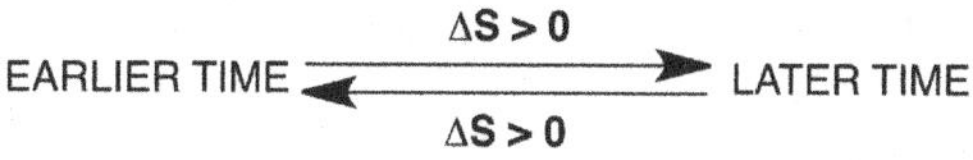

ΔS > 0
EARLIER TIME
LATER TIME
ΔS > 0

Figure 8.1 Entropy as time's double-headed arrow.

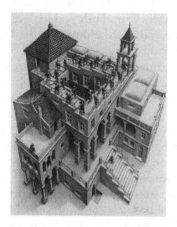

Figure 8.2 M. C. Escher's "Ascending and Descending."
© 2021 The M. C. Escher Company—The Netherlands. All rights reserved. www.mcescher.com.

point A by going uphill! Furthermore, as the Escher print shows, if you can travel uphill in *either* direction, then by continuing to travel in *one* direction, you must return to your starting point. Since thermodynamic entropy is a state function—a quantity that depends only upon the parameters of the system and not on the path taken to reach them—that would be analogous to a demonstration that the "system" of *Arcadia* is in the same state in the two time periods. Such equivalence is strongly suggested by the staging. The directions for the opening of Scene 2 (where we shift, for the first time, from 1809 to the present) read:

> The lights come up on the same room, on the same sort of morning, in the present day, as is instantly clear from the appearance of Hannah Jarvis; and from nothing else. Something needs to be said about this. The action of the play shuttles back and forth between the early nineteenth century and the present day, always in this same room. Both periods must share the state of the room, without the additions and subtractions which would normally be expected. The general appearance of the room should offend neither period ….

The landscape outside, we are told, has undergone changes. Again, what we see should neither change nor contradict.

(15)

This equilibration of the two time periods intensifies as the play proceeds; in particular, whereas the first six scenes alternate between 1809 and the present, the seventh (last) scene is set in *both* time frames (the earlier one has moved forward to 1812), and characters from both periods occupy the stage simultaneously, right up to the end. Furthermore, in the final scene the action in the contemporary period includes preparation for a costume ball, so the characters are all wearing Regency dress and are not readily distinguishable from those of the earlier period—especially the 1812 character Augustus and the modern Gus, who, the author directs, are to be played by the same actor.

What is Stoppard trying to convey by portraying in *Arcadia* a universe where changes over time are reversible—not in the Laplacean sense, but rather in the paradoxical Escherian mode of continuous change and return? I read it as a message of optimism to counter the second law's pessimistic prediction of decay. If entropy and information are mathematically related, then in Laplace's universe entropy is constant, since information is constant—everything we need to predict the future (or retrodict the past) is always at hand. As we initially took entropy and information to be *inversely* related, the creation of order seems obviously to correspond to a decrease in entropy. But as we saw earlier, Shannon's version of information theory assigned entropy and information as mathematically *identical*, not opposite: entropy increases together with information. Perhaps with that in mind, Stoppard is not much concerned with creating order—which if continued indefinitely implies yet another dismal fate the second law holds out for us: to be smothered to death by a glut of information. As philosopher Michel Serres argued: "The achievement of redundancy—when everything that needs to be said has already been said—is analogous to entropic homogeneity when matter-energy settles into terminal equilibrium" (White 1991, 268). That image finds an echo in *Arcadia*:

VALENTINE: And everything is mixing the same way, all the time, irreversibly ...

SEPTIMUS: Oh, we have time, I think.

VALENTINE: ... till there's no time left. That's what time means.

SEPTIMUS: When we have found all the mysteries and lost all the meaning, we will be alone, on an empty shore.

(94)

It should be noted that this "dialogue" involves characters from the two different time periods—they share the stage, but they are not aware of each other—one of a number of illustrations of Stoppard's deployment of theatrical technique to help convey his message. The play concludes with a couple from each era (Septimus and Thomasina; Hannah and Gus) performing a wordless waltz: an ending made poignant by our knowledge that Thomasina is to die that night in the fire.

Septimus's last phrase in the quoted excerpt can be taken to refer to entropy in *both* contexts, thermodynamics and information theory. With respect to the latter, it reminds us that information and meaning are quite different; at the same time, it alludes (I believe) to a much earlier vision of the heat death of the universe in English literature, *The Time Machine*:

> At last, more than thirty million years hence .... the red beach ... seemed lifeless again .... I saw nothing moving, on earth or sky or sea. The green slime on the rocks alone testified that life was not extinct.
>
> (Wells 1895, 199–200)

This shows, I would argue, how the pessimistic implications of the second law can be evaded in the Stoppardian (and Escherian) universe of the mind: entropy and information are always increasing but always stay the same. Perpetual motion *is* possible (see Figure 8.3 for a diagram of a perpetual-motion machine), but not as in Laplace's static universe; rather, because the ongoing creation of meaning simultaneously creates the demand for more meaning—and it is the creation, not the meaning itself, that is important. As Hannah says, arguing that none of the three research projects should be accused of triviality:

Figure 8.3 M. C. Escher's "Waterfall."

HANNAH: It's *all* trivial—your grouse, my hermit, Bernard's Byron. Comparing what we're looking for misses the point. It's wanting to know that makes us matter.

(75)

And thus Stoppard's metaphoric variant of the second law shows us how to escape its pessimistic implications—all we need is never to run out of questions. Which brings us right back to the intersection of art and science—for Victor Hugo said much the same thing, over a hundred years ago: "La science cherche le mouvement perpétuel. Elle l'a trouvé; c'est elle-même" (Lévy-Leblond 1993, 13).

Notes

1 A much shorter version of this chapter has been previously published (Labinger 1996a).
2 I have not seen productions of either of these, but have read them both. *Hapgood* certainly (on paper, at least) seems to satisfy the theatricality requirement—among other devices, there is a scene involving actors shuttling between two cubicles that invokes the "double-slit" experiment of quantum mechanics (wherein a particle can appear to pass through two openings simultaneously)—but nonetheless has never been received very favorably (Shepherd-Barr 2006, 88–89). *The Hard Problem* I found very disappointing, in part because of the gender-stereotypical portrayal of the (female) psychology student who functions as central character; but having encountered the play only on the page, I really should reserve judgment.

9 Chirality and Life[1]

As with Chapter 7, the stimulus for this study began with a chemistry paper—in this case, one that presented a particular model of how just one of the two non-identical mirror images of biochemical building blocks might have been selected for in the molecular beginnings of life. That reminded me of a literary exposition (in a rather unexpected location, as we shall see) of the role of such a process in defining what life is, as well as leading me to reflect upon the frequent appearance of the concept of autopoiesis, and its relationship to the nature of life, in L&S-related writings and presentations. Here I will discuss those three areas of inspiration in some detail, along with thoughts about connections between them.

Autopoiesis and the Nature and Origins of Life

What constitutes life, and how it came to be, has been at the forefront of both scientific and philosophical inquiry for centuries, if not millennia. As posed by Chilean biologist Humberto Maturana (one of the coiners of "autopoiesis"), the "central question" is "What is proper to living systems that had its origin when they originated, and has remained invariant since then in the successions of their generations?" (Maturana 1980, *xii*). Many attempts to define life have taken the form of identifying key properties that are characteristic of living entities, such as responsiveness to stimuli, ability to reproduce and grow, *etc.*; but many commentators have pointed out the inadequacy of such enumerations. Some things we would not want to call living respond to stimuli (*e.g.*, compass needles) or grow (*e.g.*, flames); some things we certainly *would* call living cannot reproduce (*e.g.*, mules); and so forth (Luisi 2016, 120–121).

In 1979 Maturana and co-author Francisco Varela published an important essay that argued for a more systems-based approach, and in doing so invented a neologism, autopoiesis—*poiesis* from the Greek for making—whose origin itself had some literary inspiration:

> [W]hile talking with a friend ... about an essay ... in which he analyzed Don Quixote's dilemma of whether to follow the path of

DOI: 10.4324/9781003197188-9

arms (*praxis*, action) or the path of letters (*poiesis*, creation, production) ... I understood for the first time the power of the word "poiesis" and invented the word that we needed: *autopoiesis*. This was a word without a history, a word that could directly mean what takes place in the dynamics of the autonomy proper to living systems.

(Maturana 1980, *xvii*)

Whether the word actually lacks any history might be questioned—Bernal used the related "biopoiesis" in his lengthy examination of the origin of life (1967, 54)—but both the term and the concept have proven highly fruitful for subsequent scholarship.

What is meant by autopoiesis? Clarke provides a concise and useful definition (Maturana and Varela's language, it must be acknowledged, is not always as transparent as one might like):

Autopoiesis—literally, "self-making"—named the recognition that a living system, such as a cell or an organism built up from cells, is a self-referential system: it is the processual product of its own production .... The autopoietic process turns upon itself, recursively: the organization enables the production that maintains the organization, and so on.

(Clarke 2011, 222)

Crucially, the focus is not on the parts that constitute a cell or organism, but rather, on the *interactions* between those parts:

[O]ur problem is the living organization and therefore our interest will not be in properties of components, but in processes and relations between processes realized through components ... we are emphasizing that a living system is defined by its organization and, hence, that it can be explained as any organization is explained, that is, in terms of relations, not of component properties.

(Maturana and Varela 1980, 75–76)

That the emphasis *must* be so placed can be seen by considering the chemicals that make up living cells, and the means by which they are produced and mobilized in maintaining cellular function. The major components of a cell include proteins, which are chains of amino acids; nucleic acids, which are chains of nucleotides; cell membranes, composed of lipids (fatty acids); and a (large) number of other chemical species. These are variously synthesized within the cell and/or imported from the outside environment. A living system must be *bounded*, as by a membrane in the case of a cell; but it cannot be a *closed* system, a term that is sometimes (erroneously) applied. There has to be exchange of material and energy with the environment, across the boundary, in order to sustain the processes that comprise life; because as we saw in Chapter 8, a closed system must sooner or later succumb to the second law of thermodynamics.

However, the generation of each of those components is inextricably bound to the presence of all the others. Assembly of amino acids into proteins is governed by the genetic code of DNA and effectuated (in part) by the transcription of DNA into RNA and subsequent translation of the latter; but that translation process depends upon enzymes, which are proteins. The modified lipids that constitute the membrane are synthesized with the aid of other enzymes; but those transformations must take place within the bounded cell, or else they would diffuse away and be unavailable for assembly. And so forth. But the "control" mechanism for all of this activity resides entirely within the cell—it manages its own management—which distinguishes such an autopoietic system from its non-living counterparts.

Subsequently there has been a tendency to generalize autopoiesis to apply to a wide range of areas, including some that are not obviously relevant to living organisms: social organization, textual analysis, and others (Clarke 2011, 222). That seems a questionable move, insofar as the concept was explicitly introduced to distinguish living and non-living systems. Indeed, Varela and Maturana themselves expressed skepticism (Luisi 2016, 151), while others offered much stronger objections:

> Autopoiesis was introduced into the literature by Maturana and Varela as the name for a particular system description which they claimed was necessary and sufficient to define the living and also to explain it. The term has been widely applied in the literature instead to spontaneous order production or self-organization in general, whether living or not … this particular move must be rejected … if the concept of autopoiesis can be used this way it immediately shows the concept's failure to define and explain the living, making it enigmatic as to what is being generalized.
>
> (Swenson 1992, 207)

Without necessarily endorsing such linguistic puritanism, I think use of the alternative, more general term "systems theory" is much more appropriate for these extended applications. Detailed exploration of the latter would take us too far afield from the main point of this chapter; Clarke's brief essay on the topic (2011) provides a good introduction and useful further references.

The significance of autopoiesis to the *beginnings* of life is less clear, although it certainly seems reasonable that there *should* be a connection:

> Autopoiesis is not directly concerned with origin of life; rather, it is an analysis of the living as it is—here and now, as the authors say.

We will see, however, that once the question "what is life?" is clarified by the theory of autopoiesis, we will receive an indication on how to proceed in tackling more properly the question of the origin of life.

(Luisi 2016, 123–124)

As noted earlier, many scientists (and philosophers, and others) have tackled that question, usually without much, if any, reference to autopoiesis. (Luisi himself does not offer much more than a brief suggestion along those lines, as we shall see in a moment.) Most proposals (leaving aside creation myths) begin with the generation of a collection of moderately complex organic chemicals: some of the previously mentioned cellular components such as amino acids, nucleotide precursors, *etc.* Many studies, going back to the classic 1950s work of Stanley Miller, have demonstrated the synthesis of such compounds from much simpler building blocks (such as hydrogen, ammonia, methane, water, and cyanide) under conditions understood to be characteristic of the prebiotic world (Luisi 2016, 39–55). An alternative idea is that life was "imported" from elsewhere in the universe—the "panspermia" theory—supported (somewhat) by the common occurrence of such organic compounds in meteorites. However, that doesn't solve the problem of how life developed from these substances; it merely moves it to another location.

Once the building blocks have been obtained, they must be assembled into entities that satisfy the minimal criteria of life. As we saw previously, living processes are maintained by many components interacting with one another; how could such complexity evolve spontaneously? I do not have space to discuss the wide range of chemical possibilities, some extremely detailed, that have been considered (see Bernal 1967; Luisi 2016; Marshall 2020, among many other accounts), but I will briefly distinguish two global views. One is the "RNA world" theory, which is based on the fact that RNA can exhibit the sort of catalytic function (so-called ribozymes) more typical of proteins. Conceivably, then, RNA could have been the crucial prebiotic substance, capable of carrying out many of the functions requisite for life, with subsequent evolution of the more complex multicomponent systems we have now (Dyson 1999, 12–14). An opposing view is that, since more advanced experiments of the Miller type have shown that essentially *all* of those key components could have been produced simultaneously by prebiotic chemistry, spontaneous organization of much of the whole complex mix may not be so unreasonable after all (Powner and Sutherland 2011). Luisi believes that the latter scenario is more consistent with the ideas of autopoiesis (401–402). Active research along both of these lines (and others) continues, although nobody has yet achieved even a pale approximation of a living cell under laboratory conditions.

Chirality and Its Role in Life

The definition and structural basis of chirality are presented in more detail in the Appendix; briefly, a molecule with four different groups attached to a central carbon can be *chiral*. That means it is not identical to its mirror image; the two mirror images are called *enantiomers*. They can be distinguished by the fact that they rotate polarized light in opposite directions, a property known as *optical activity*. They will also interact differently with other chiral entities, but *not* with achiral ones. This is easiest to envision by considering macroscopic chiral objects, such as hands. The fit between your right hand and your neighbor's left and right hand is clearly different, as we recognize every time we shake hands. In contrast, there is no difference between how right and left hands lie on an *achiral* object, such as a table.

It follows that if we synthesize a chiral compound under completely achiral conditions—not under the influence of any agent that can discriminate handedness—there will be no reason to favor one enantiomer over the other, and therefore we should obtain equal amounts of the two, which we call a *racemic* mixture. Thus, for example, the prebiotic generation of amino acids, which are chiral, should—one would think—have yielded strictly racemic products, since there should have been no chiral environment. What we find, though, is that *all* amino acids in *all* living organisms are found only as a single enantiomer (Figure 9.1)! The same is true of essentially all chiral species in living systems; this is such a universal characteristic of life (as we know it) that an upcoming (2022) mission to Mars designed to search for evidence of past and/or present life will be equipped with a robotic laboratory capable of detecting the presence of any predominant enantiomeric forms of molecules (Remmel 2020, 34).

The reasons for that are not hard to understand, since the interaction between molecules is crucial to metabolism and, hence, to life. The way two chiral molecules interact depends strongly on their geometric

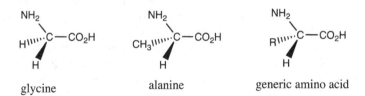

glycine   alanine   generic amino acid

Figure 9.1 Structures of the two simplest amino acids and a generic one; there are around 20 such used in proteins, with different substituents R. Note that all have the same configuration; only glycine is achiral, having two of the substituents on the central carbon the same (H).

relationships. If we label the two enantiomers of a given molecule as R and S (there are other conventions of nomenclature, but this one is convenient), then the way the R form of molecule 1 fits with the S form of molecule 2 will be significantly different than that of R-1 with R-2, and so on; as a consequence, they will generally exhibit different biochemical behaviors. A well-known example is that of thalidomide, which is a chiral substance. Initially the drug, intended for use (among others) by pregnant women suffering from morning sickness, was distributed as the racemic mixture, with the well-known disastrous consequences. Too late, it was realized that only the R-isomer is a tranquilizer, while the S-isomer is a potent teratogen.

With respect to the *origins* of chirality, there is an obvious bootstrap issue. Since enantiomers are differentiable only by virtue of their interactions with other enantiomerically pure substances, how did any preference get established in the beginning? That development, however it took place, must have played an important role in the establishment of life. I will first look at a discussion of that question in a work of literature—detective fiction, of all things.

The Documents in the Case is a mystery novel, published in 1930, by Dorothy Sayers (we encountered her well-known protagonist, Lord Peter Wimsey, in Chapter 8; he does not figure in this work) and Robert Eustace. A brief summary: two former schoolmates, an artist (Lathom) and an author (Munting), take up lodging in a house occupied by an older retired engineer (Harrison) and his much younger second wife. Lathom and Mrs. Harrison start an affair, unknown to Harrison, who is an avid mushroom hunter and regularly goes off to an isolated shack to indulge in his passion. On one such occasion Lathom accompanies him, but leaves for a couple of days, and returns (with Munting in tow) to find Harrison dead, apparently from eating the wrong kind of mushroom. Indeed, examination shows he succumbed to poisoning by muscarine, the active agent in *Amanita muscaria*, which sufficiently resembles the edible *A. rubescens* to make it plausible (to some) that he had made a mistake in harvesting.

The action is expounded somewhat atypically for a detective novel: partly in epistolary format and partly in the form of dual first-person narratives by Munting and Harrison's son. The latter can't believe that his father could have made such an error, and when he discovers evidence of the love affair in some of the letters, he is convinced that Lathom murdered his father somehow, and induces Munting to help his investigation. They learn that Lathom had had access to a chemistry laboratory containing synthetic preparations of a number of natural products—including muscarine—strongly suggesting the means of the crime, but agree that (as far as they know) there would be no way to prove it.

Subsequently, Munting attends a dinner gathering with a chemist (Waters), a physicist, a biologist, and a clergyman; and the talk turns to one of his pet interests:

"What is Life?" I asked, suddenly.

"Well, Pontius," said Waters, "if we could answer that question we should probably not need to ask the others. At present—chemically speaking—the nearest definition I can produce is that it is a kind of bias—a lop-sidedness, so to speak .... That is to say, up to the present, it is only living substance that has found the trick of transforming a symmetric, optically inactive compound into a single, asymmetric, optically active compound. At the moment that Life appeared on the planet, something happened to the molecular structure of things. They got a twist, which nobody has ever succeeded in reproducing mechanically—at least, not without an exercise of deliberate selective intelligence, which is also, as I suppose you'll allow, a manifestation of Life."

(244)

Waters goes on to explain how synthetic and natural substances may be distinguishable by rotation of polarized light, as described in Appendix 1. Munting recognizes (*very* reluctantly) that Lathom's guilt might be proven thereby; the experiment is carried out; the muscarine found in Harrison's body is indeed optically inactive; and Lathom is duly convicted and executed.

Of course, this *is* a detective story, and a prime reason for the science is to move the plot forward; but there is also a substantial amount of deeper thinking. The chemical knowledge is presented at a sophisticated level[2] and (mostly[3]) accurately:

The substance produced by synthesis always appears in what is called a racemic form. It consists of two sets of substances—one set having its asymmetry right-handed and the other left-handed, so that the product as a whole behaves like an inorganic, symmetric compound; that is, its two asymmetrics cancel one another out, and the product is inactive and has no power to rotate the beam of polarized light.

(245)

and other topics such as chemical evolution and the second law of thermodynamics are brought up as well. The central importance of chirality to life is likewise clearly expounded, except perhaps with respect to the timing: it is unlikely that enantiopure substances originated "at the moment that Life appeared on the planet"; they were more probably a subsequent development. That leads us to our final topic: the origin of chirality.

How Could Chirality Have Arisen?

Sayers and Eustace observe that enantiomeric selection is a characteristic of life, but life is requisite for selection. How then could it all have started? As with the organization of living species from precursors, there are two basic approaches to this problem. One considers that there is some *inherent* asymmetric force in the universe, such that a preference—even though infinitesimally small—could have been present *ab initio* and somehow amplified over the eons. Indeed, it has been known since the 1950s that the so-called weak force of atomic nuclear interactions *is* asymmetric, and some have proposed that this could result in a preference for one enantiomer over the other. The alternative is that the emergence of chirality was entirely stochastic, the result of random effects giving a tiny imbalance that was likewise gradually amplified (Meierhenrich 2008, 5–16).[4]

We will focus on the second case: how could what starts as a perfectly equal distribution of two enantiomers spontaneously, of its own accord, begin to prefer one? Most of the proposed explanations are centered on the concept of *autocatalysis*: the ability of a species to function as a catalyst for synthesizing more of itself out of precursors. One model runs as follows: suppose compound A reacts with B to give X, and the reaction is accelerated by the presence of X acting as a catalyst. A species acts as a catalyst by interacting with the reacting molecules, and as emphasized on p. 152, such interactions are crucially dependent on geometric fit, which depends (*inter alia*) on chirality. Hence, if X is chiral, we could imagine one enantiomer of X (call it X_R) preferentially catalyzing formation of that same enantiomer. But that by itself doesn't get us anywhere. In the achiral prebiotic environment, the same would happen for both enantiomers—X_S would catalyze formation of more of itself—so if the starting distribution is racemic—50:50—it will be maintained as the population grows.

Now let us tweak that scenario a little, and propose that the catalytic species are composed of *two* units of X (X-X dimers). Note, as previously remarked, that the enantiomers of X will interact with one another in different ways: we can have homodimers X_R-X_R and X_S-X_S as well as the heterodimer X_R-X_S. Suppose further that only the homodimers can function as catalysts, such that X_R-X_R and X_S-X_S catalyze formation of more X_R and X_S, respectively, while X_R-X_S is inactive. Such a model, as I will show in a moment, provides a mechanism for amplifying a tiny imbalance between enantiomers, in principle ultimately leading to a complete preference for one over the other.

But could even that tiny initial imbalance have arisen in a purely achiral environment? In fact it is not only possible but expected, by the laws of statistics. Consider a coin toss, which (like the abiological synthesis of a chiral product) has two equally probable outcomes. If we do *enough* trials, we confidently expect an equal number of heads and tails to result. But that does not mean that we would expect any *single* trial of any limited size to come out that way. As the simplest case, take sets of two tosses. Probability theory (and common sense) tells us that we will get two heads or two tails each for one-quarter of the sets, and one head, one tail for half. In other words, half the sets will *not* give a 50:50 result! If we do *many* such trials and add up all the results, the total will approach half heads, half tails; but that is *not* the case for just one set.

As we increase the size of a set, of course, the expected deviation from 50:50 decreases. For 100 tosses, the average expected outcome works out to a 53:47 distribution. The same applies to molecules: if we generated an assembly of 100 chiral molecules by a completely achiral process, we would expect an average composition of 53 of one enantiomer and 47 of the other—*not* a perfectly racemic mixture! If we could measure the optical rotation of such a small sample (we can't) we would expect a non-zero value. Again, if we repeated the experiment many times, we would expect the sets to exhibit a range of distributions, with equal numbers of positive and negative rotations (as there is no inherent preference for one or the other), adding up to zero. But probability tells us that *any one* such synthesis will most likely give a result that differs from zero.

It is useful to introduce here a quantity called *enantiomeric excess* or ee, the fractional preference for one enantiomer over the other, which is mathematically defined as $(R-S)/(R+S)$, where R and S are the numbers of molecules of X_R and X_S, respectively. So the average ee expected for the 100 molecule set would be $(53–47)/(53+47)$ or 6%. What happens when we expand to a physically significant quantity of the substance, say a gram? One gram of a typical biological molecule consists of around 10^{20} molecules. The ee expected for a purely statistical outcome would be miniscule, around 10^{-10}, many orders of magnitude below detectability.

But now look at what happens under the dimer-based autocatalytic scenario outlined on p. 154. It's easiest to see how the system evolves if we start with a very small assembly of molecules (a larger initial state eventually ends up the same way). Suppose we have somehow generated 200 molecules in all. As discussed earlier, even in the absence of any chiral influence, obtaining exactly 100 of each enantiomer is *not* the most likely situation. Suppose further, then, that the sample is comprised of 101 X_R and 99 X_S—which would be well within the statistically expected range of probable imbalance for a sample of this size—giving an ee of 2/200 or 1%. Allow them to dimerize, assuming that there is no preference for homo- *vs.* heterodimers. (If there is a preference, as there generally will be, the numbers will be different; but the basic process works in the same way.) The 200 monomers will form 100 dimer molecules, and since there is slightly more X_R in the pool, the most probable relative proportions of X_R-X_R, X_R-X_S and X_S-X_S formed will be 26:49:25.

Now let the synthetic reaction—A + B to give X, catalyzed by (homo) dimers of X—proceed. For convenience of counting, let it run for two rounds—that is, each X_R-X_R and X_S-X_S catalyzes formation of two more copies of X_R and X_S, respectively, while X_R-X_S just sits there. Then we will have made 52 more molecules of X_R and 50 of X_S. Adding these to the starting population, we now have 153 X_R and 149 X_S, with ee

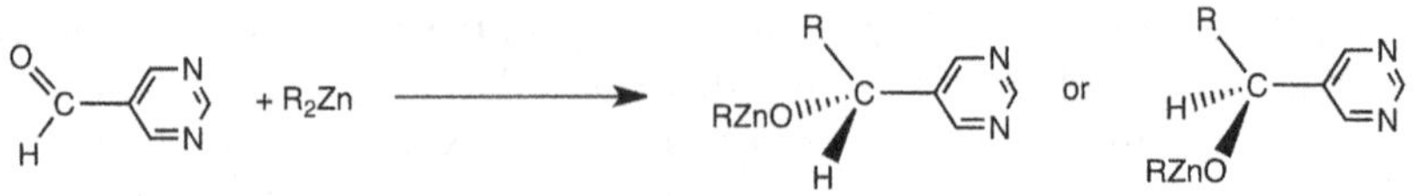

Figure 9.2 Schematic representation of the Soai reaction. Organozinc reagent R_2Zn adds to the C=O double bond to give either of the two enantiomers of the product shown; each enantiomer serves as a catalyst to accelerate its own formation.

now $4/302 = 1.32\%$—higher than what we started with! The resulting dimer population will be $X_R\text{-}X_R:X_R\text{-}X_S:X_S\text{-}X_S = 39:75:37$; the next round of catalyzed reaction will end with 231 X_R and 223 X_S, or ee = 1.76%; and so on. It should be clear, then, that if this process continues long enough we can end up with a macroscopic sample of essentially 100% ee—enantiomerically pure! Note that *which* enantiomer was preferred is purely a matter of chance: if we ran this scenario many times we should enrich either of the two enantiomers with equal probability.

Is this just a mathematical fantasy? No: such autocatalytic enantiomeric enrichment has been demonstrated experimentally! The reaction used is shown in Figure 9.2, where the product serves to catalyze its formation from the two reactants. In one set of experiments a small amount of product pre-prepared with a small excess (ee = 2%) of the S enantiomer was added at the outset, and sequential runs carried out, with the product from one being used as catalyst for the next. As can be seen (Figure 9.3), the ee was *rapidly* amplified, from 2% to 88% in only four rounds (Soai *et al.* 2000). Even more dramatically, a second set of experiments was performed similarly but using a completely racemic preparation of product/catalyst to start. The outcome of 37 such experiments was an average final ee of around 70%, favoring S 19 times and R 18 times—just what would be expected for a randomly based process (Soai *et al.* 2003).

Results such as these provide strong evidence for the ability of *some* autocatalytic mechanism to generate enantiopure species out of racemic starting materials, but not necessarily for the specific one considered. Such evidence *was* subsequently reported in the paper that was my original inspiration for this chapter. The details are too complex to present, but the basic finding was that the rate data for the reaction under a range of conditions could *only* be fit to the model described here, in which homodimers are the catalysts and heterodimers are inactive (Blackmond *et al.* 2001).

But that, of course, only validates the model for the specific chemistry under investigation; it is far from proof that it was operative during prebiotic times. Indeed, a number of alternate proposals have been put forward, some with experimental support (Meierhenrich 2008, 5–16). I will

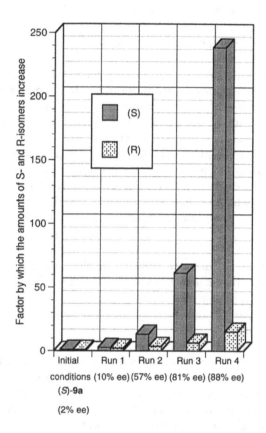

Figure 9.3 Data showing how a very low initial ee is rapidly amplified during successive runs of the reaction shown in Figure 9.2.

Reprinted from Soai *et al.* 2000. Copyright American Chemical Society.

not examine them here, except to return briefly to the "non-random" hypothesis: that some universal property induced a small preference for one enantiomer over another. There is (at least) one argument in its favor. As mentioned on p. 150, a number of biochemical building blocks, including amino acids, are found in meteorites, and some of those samples have been found to be non-racemic, usually with low (a few percent) ee values (Meierhenrich 2008, 146–148). Strikingly, when that is the case, the favored enantiomer is *always* the same one used on earth in living systems! It's hard to see how that could be just a coincidence, as the mechanism we've discussed for abiological amplification starting with strictly racemic materials has an equal probability of ending up at either of the two enantiopure endpoints. There have been tests of some models for initial bias, but no clear support. For example, the degree of preference that would be imposed by the asymmetry in the weak force has been shown to be far too small to perturb the outcome of the Soai system (Hawbaker and Blackmond 2019).

Luisi is agnostic on this point, posing it as a question:

> The origin of homochirality in nature is usually debated in terms of two opposing views. According to a deterministic (*ex-lege*) approach, one of the two enantiomers has a lower intrinsic energy and therefore a greater probability of occurrence. The alternative is a stochastic process, according to which the selection of one enantiomer over the other out of a racemate was determined by contingency. From recent experiments, it appears that the breaking of symmetry may be achieved rather easily in the laboratory, possibly under prebiotic conditions. Do you agree then with the view, that the origin of homochirality in nature "is no longer a problem"?
>
> (Luisi 2016, 406–407)

I suppose one could call it "no longer a problem" in the sense that we now have multiple models that *could* account for it without invoking anything highly implausible, but we don't seem to be close to figuring out how it *actually* happened. But perhaps that's beyond any point we might reasonably expect to reach.

Does autopoiesis have anything to say about this issue? It may seem that a random model—a system in which everything guiding the selection of a single enantiomer comes from *within* the system—is much more consistent with the tenets of autopoiesis than one in which a preference, however small, is imposed from the outside. However, of course, that does not constitute any sort of valid argument for such a mechanism, since we have no grounds for assuming that autopoiesis is relevant to the *origins* of life in the same way that it describes the *maintenance* of life. Hayles comments that autopoietic processes are "always operating in the present moment" (1999, 139) and that "evolution is a sore spot for autopoietic theory" (149). That suggests, perhaps, that one should *not* try to draw any inferences about origins from autopoietic thinking.

Conversely, though, might these studies of how chirality may have originated have something to say about autopoiesis, or systems theory more generally? This falls into the realm of speculation, but some commentators do suggest at least a possible connection. Maturana speaks of "structural coupling" of "independent systems that, by triggering in each other a structural change, select in each other a structural change" (1980, *xx*). That sounds very much like what is going on in the previously described model: two independent molecules (systems) couple to form alternate structures (homo- or heterodimers), which leads to selection of the next structure to be formed by their catalytic action.

Similarly, Hayles paraphrases Maturana's treatment of the origins of language (about which she is dubious, it must be noted) as involving "a process whereby observers ... provide the triggers that help other observers similarly orient themselves within their domains" (147). A related comment comes from an explication of Maturana's theories of autopoiesis and cognition:

> Within this context, Maturana distinguishes two types of interactions between organisms. The first is where the behavior of one leads directly to the behavior of the other, for example, fight/flight or courtship. The second is less direct. The behavior of the first organism "orients" a second organism, i.e., directs its attention to some other interaction that the two have in common. The orienting behavior stands for or represents something other than itself. What is important is that the behavior symbolizes something other than itself, and its success depends on the common cognitive domains of the organisms. This leads Maturana to describe the domain of such behaviors as *consensual*, and the interactions as communication.
>
> (Mingers 1991, 325)

Is it too much of a stretch to read that as evocative of the autocatalytic model for homochirality we discussed? We could describe the latter (at the risk of getting overly anthropomorphic) as involving an "organism" or an "observer"—here it would be a molecule—interacting with a second molecule with which it has something in common (the same handedness) and thus "orienting" it to produce a successful behavior: the ability to act as a catalyst to produce more of itself. Conversely, the interaction with a molecule of the opposite handedness—one with which the first does *not* have something in common—is inhibitory: the resulting heterodimer is catalytically inactive.

Another example comes from an article on systems theory and literature, which includes a gloss on Luhmann's gloss on another author (George Spencer-Brown):

> The world makes itself distinct from itself by means of a distinction .... To observe is to draw a distinction ... but where, after having drawn a distinction, are we to place the observer?... To decide where he belongs requires a second observer.... That is, the world marked or indicated by the first distinction is identical with the observer. Only a second distinction can distinguish the first distinction and thereby the observer who uses it.
>
> (Roberts 1999, 28–29)

This passage is admittedly more than a little opaque, but again there are aspects that seem to resonate with our chirality model. A single observer (a single molecule of catalyst) cannot distinguish the world from himself (cannot bring about the amplification needed to convert a racemic substance into a homochiral one); only when a second observer (a second molecule) is brought in can that be achieved.

All this may be stretching analogy too far: it is quite possible that these nominal similarities of language are coincidental. Nonetheless, I think one could imagine a connection that is not entirely coincidence, such that mechanistic and perhaps even *quantitative* aspects of the model in chemistry might be brought to bear on considerations of how very different sorts of systems—cognitive, social, *etc.*—evolve. However, that is a project for the future.

Notes

1 This chapter is based on a talk presented at the 2004 (Durham NC) SLSA meeting.
2 The source of their chemical knowledge has been traced (Manchester 2007) to a well-known (at the time) lecture on "Stereochemistry and Vitalism" by chemistry professor Francis Japp (1898); some of the language employed by Sayers and Eustace closely tracks that of Japp.
3 Unfortunately, the accuracy didn't extend as far as the more detailed chemistry. The formula for muscarine is given as $C_5H_{15}NO_3$, and the procedure for its synthesis outlined: "[Y]ou can make it by heating ethene oxide with triethylamine. That gives you your cholin. Then you oxidize it with dilute nitric acid—the stuff you etch with, you know. Result, muscarine" (Sayers and Eustace 1930, 232). That description may be reformulated as an equation (Figure 9.4).

   Alas, there are several things wrong. The intermediate is not choline (as we now spell it); the formula of the product would be $C_8H_{17}NO_2$, not what was given; and neither the formula nor the structure for muscarine is correct (in fairness to the authors, the actual structure was not conclusively known at the time). Most importantly, though, the product implied by the synthetic procedure could *not* have been optically active, having no carbon center with four different substituents. The actual structure of muscarine *is* asymmetric—it does exist in optical isomers—so the central idea was still basically correct; except that we also now know that muscarine is only a very minor component of *A. muscaria*, whose toxicity—which is not all that high compared with other poisonous mushrooms, such as *A. phalloides*—is primarily due to other components. See https://en.wikipedia.org/wiki/Amanita_muscaria.

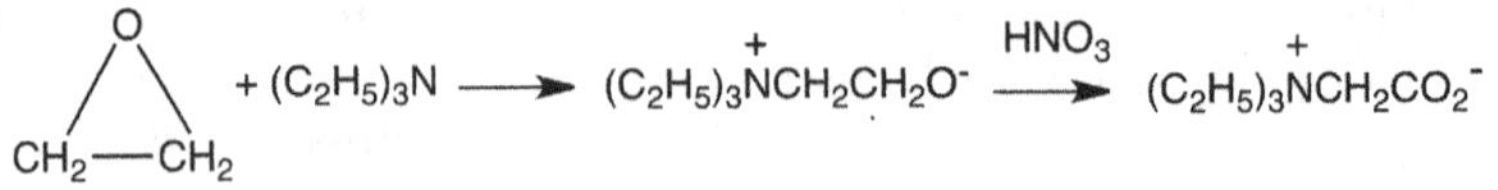

Figure 9.4 Chemical equation corresponding to the synthesis of muscarine as outlined by Sayers and Eustace.

4 Primo Levi discusses this question in an essay written for the non-technical reader (Levi [1984] 2018), noting that stereochemistry was the topic of his dissertation. He speculates on several possible sources of a fundamental universal asymmetry, but does not seriously consider a purely probabilistic initial event, and offers no suggestions for an amplification mechanism. I thank Rebecca Falkoff for calling this essay, and a couple of related pieces, to my attention; see also Falkoff (2021).

10 Making New Life[1]

In 2003 I attended a presentation by David Baltimore, then president of Caltech, with the title "Biotechnology is the new, new thing." He talked primarily about practical and economic aspects of the topic (the talk was aimed at an entrepreneurial organization), but also spent some time speculating on the future of "synthetic biology" and such matters.

Around the same time I ran across a comment from a literary critic about the potential value of relating biotechnology to the humanities:

> New technologies in medicine and genetics ... give human beings unprecedented power to act on the conditions and habitats of life itself, to remake what once seemed a nature beyond human control .... The world of nature and the world of human making, conceptually separated since the Enlightenment, are thus bound together more strongly and in more ways than ever.
>
> (Paulson 2001, 2)

It occurred to me that over the last couple of years I had read—almost entirely by chance—a number of books in which the creation of new forms of life played an important part. Some involved generation of new or modified beings via advanced methods of genetic manipulation, while others hearkened back to earlier traditions. An exploration of whether and how the various authors might have been informed by recent developments in the field seemed worthwhile.

I begin this chapter with brief historical sketches of biotech and genetic manipulation relevant to modification and/or creation of life (including CRISPR, a recent development that appears to offer breakthrough capabilities), as well as literary manifestations thereof. I will then look in more detail at the particular books that attracted my interest, and address several themes that are prominent in most or all of them: the incorporation of biotech developments; the possibility of control; and the interplay between language and science.

DOI: 10.4324/9781003197188-10

A (Highly Selective) History of New Life in Biology and Literature

In biology, we might choose any of a number of plausible starting points; one I like is Linus Pauling's description of sickle cell anemia as "a molecular disease." That arose from his finding that hemoglobins isolated from normal and diseased red blood cells exhibited different mobilities in an electric field, showing that they differed in ionic charge by a small amount. He proposed that was a consequence of altering just a couple of amino acid residues per (very large) protein molecule, which affected the way the latter interacted with one another, thus perturbing the shape and oxygen affinity of the cells (Pauling *et al.* 1949). That such a subtle change could have such a drastic effect seemed remarkable; but it was confirmed a few years later, when it was reported that the sequences of normal and sickle cell hemoglobins indeed differ by exactly *one* amino acid out of a chain of nearly 300: a charged residue (glutamate) in the former is replaced by an uncharged one (valine) in the latter (Ingram 1957). The implication—not realized at the time, of course, since the techniques were not yet available—is that properties of life could be *significantly* modified by relatively minor "editing" of the genes responsible for determining protein sequences.

By this time it was already known, primarily from the work of Avery and coworkers (1944), that DNA, not proteins themselves, was responsible for carrying genetic information. A key event was the elucidation of the structure of DNA, which pointed towards the mechanism by which information is transmitted during replication (Watson and Crick 1953). Understanding how triads of DNA bases encode for individual amino acids in proteins, along with the mechanism by which that is accomplished, was achieved over the ensuing decade or so (Divan and Royds 2016, 7–12). The ability to incorporate non-native "recombinant" DNA into organisms, as well as methodology for sequencing and synthesizing DNA strands, was introduced in the 1970s; the latter two were steadily improved *via* automation to the point where they can be done extremely quickly and cheaply today (Davies 2018, 28–36).

Molecular biological research has thus built up a genetic "toolbox" that makes it possible to take an organism and either modify existing genes or insert new ones to obtain GMOs—genetically modified organisms. This is the realm of synthetic biology, which has been defined as

> the creation of new living systems by deliberate design … rang[ing] from the modification of existing organisms to do entirely new things, which is now routine at least on a small scale, to the as-yet unrealized creation of a living organism from non-living components.
>
> (Davies 2018, 2–3)

To be sure, such capabilities are still *far* from limitless, because of not only technical problems but also ethical concerns. A conference on potential problems associated with recombinant DNA, held in Asilomar in 1975—shortly after the methodology was developed—recommended caution (Berg *et al.* 1975). Guidelines for its safe use have generally been effective over the ensuing years, but concerns remain; societal resistance to GMOs in foods, for example, remains strong. One case in point is the development of "Golden Rice." The insertion of a gene for production of beta-carotene into ordinary rice was accomplished in the 1990s, with the goal of reducing vitamin A deficiency–related diseases among populations whose diets are heavily dependent on rice (Davies 2018, 17–19); but it took over 20 years before it was approved (tentatively) for use in Bangladesh (Stokstad 2019).

A particular concern is that much of this methodology is *too* easy, leading to significant DIY or "biohacker" activity (Baumgartner 2018). Fears that someone might inadvertently—or even deliberately—create a virulent microorganism, or an invasive plant or animal species that could escape from the laboratory into the world at large, are widespread; this is a theme of (at least) one recent novel, Richard Powers' *Orfeo* (2014). The development of CRISPR, a new, possibly breakthrough technique, has considerably magnified the future of biotech, in terms of both opportunities for positive applications and risks of negative outcomes (Doudna and Sternberg 2017). Given its potential importance (recognized, *inter alia*, by the 2020 Nobel Prize in Chemistry to two of its inventors, Jennifer Doudna and Emmanuelle Charpentier), it is worth spending a little time here on a brief description.

CRISPR exploits the discovery of a defense system found in certain bacteria, whose DNA contains stretches of "clustered regularly interspaced short palindromic repeats"—hence the acronym—that match up with DNA from hostile viruses known as bacteriophages. When a virus invades, the bacterium transcribes those stretches to produce pieces of RNA that bind to the corresponding stretches of the viral genome, and thereby guide a nuclease called Cas9—an enzyme that cuts DNA strands—to sever the gene and thus inactivate the virus. In principle, then, one can design and synthesize an analogous construct to target any desired DNA sequence in any organism. Of course, simply cutting such a sequence will rarely be of any use. But many organisms have repair systems, some of which operate by detecting a break in the strand of DNA, finding a template that has a sequence identical to that on both sides of the break, and using that to repair the damaged site. If one adds yet another construct, consisting of a *new* sequence flanked by the matching stretches on the sides of the break, then the repair system can (it doesn't always) insert that new sequence instead of the original excised one, resulting in a "revised" gene and hence, to some degree, a new organism (Davies 2018, 133–135). Further advances in the method's reliability and versatility continue to be developed (Anzalone *et al.* 2019).

While all this may sound pretty complicated, in practice it is relatively straightforward, cheap, and hence potentially highly powerful compared with earlier genetic editing methods. One milestone was recently reported: an apparent "cure" for sickle cell anemia, the condition that introduced this section. CRISPR was used not to correct the single mistake in the hemoglobin gene, but to edit a different gene that enabled production of fetal hemoglobin in the patient's blood stem cells (Brainard 2019). Much more controversial is making changes in germ line cells, which can be passed on to subsequent generations. In 2018 a scientist in China performed such editing on embryos to eliminate susceptibility to HIV, and then implanted them into a woman (whose partner was HIV positive), eliciting virtually universal outrage (Cyranoski and Ledford 2018). Another potentially dramatic move, still only at the talking stage, might be eliminating the transmission of malaria by mosquitoes by means of a CRISPR-based "gene drive" (James 2005). But many are nervous, even skeptical; one biologist commented: "Nothing in history suggests that those who control and profit from material production can really be depended upon to devote the needed foresight, creativity, and energy to protect us from the possible negative effects of synthetic biology" (Lewontin 2014). Proper consideration of the ethical and ecological implications of such global transformations is still at an early stage.

In literature, the creation of new forms of life has been a common theme for centuries, whether invoking supernatural forces or the latest scientific developments. For the former, we can go back to the Jewish legend of the golem, which has taken various forms over the years. One is that of the golem of Prague, created out of clay by a 16th-century rabbi to protect the Jews against pogroms. Magical incantations were generally required to animate the golem; but in one version—with particular literary relevance?— it was accomplished by means of *text*. Writing the three Hebrew letters *aleph*, *mem*, *tav*—spelling *emet* (אמת) or truth—on the creature's forehead brought it to life. After it was no longer needed—or, more frequently, when it ran amok and caused havoc—it could be killed by erasing the *aleph*, leaving *met*, which means dead (Oreck n.d.).

By far the most familiar creature is that in Mary Shelley's *Frankenstein* (1818), which almost immediately became, and has remained, deeply embedded in Western culture. Whether Shelley was inspired by—or even aware of—the golem legend is debatable (Simon 2018). There are some obvious similarities, but at least one clear difference: Frankenstein's creation was animated not by magic but by science. Shelley (along with her husband-to-be Percy) was interested in, and well up on, the latest scientific developments in chemistry, electricity, *etc.* (Robinson 2017); and her protagonist, Victor Frankenstein, is converted by one of his professors to give up his youthful unhealthy fascination with the alchemists in favor of modern chemistry (Shelley [1818] 2017, 30–31). In fact, Shelley does not give us *any* actual science: the universally recognizable image of the creature being brought to life by electricity comes from the 1931 film, not the novel itself. She was a little more explicit about the possibility of re-animating a corpse *via* "galvanism" in her introduction to a later edition (Shelley [1831] 2017, 192). But the description of the animation process in the original novel itself—both of the original creature and of an aborted attempt to provide it with a mate—is given in completely general terms (41, 138); only the phrase "that I might infuse a spark of being into the lifeless thing" (41) suggests the involvement of electricity.

Indeed, far more attention has been paid to ethical than scientific aspects of the work; Victor has been condemned more for abandoning his creature than for the hubris of making it. Many of the worries about new biotech methodology refer back to *Frankenstein* as the original cautionary tale. Nonetheless, it is clear that Shelley fully intended to shift the realm of life creation narratives from the supernatural to the natural, a move that has dominated over the ensuing two centuries. For a good while, aside from the virtually innumerable works in various media (film, stage, novels, comic books, *etc.*) that took direct inspiration from Shelley, the new life forms were more typically not based on organic matter, but were entirely artificial—robots. The word itself was introduced in Čapek's 1920 play *R.U.R.* (for *Rossumovi Univerzální Roboti* or Rossum's Universal Robots), although there were many earlier examples (*e.g.*, the mechanical doll Coppelia in E. T. A. Hoffmann's story, as well as an opera (Offenbach) and ballet (Delibes) based thereon; Tik-Tok in Baum's Oz series; *etc.*). Of course the concept has featured still more prominently in 20th-century science fiction, most familiarly in the work of Isaac Asimov.

Literary exploration of life creation *via* biological methods evolved somewhat more slowly through the 20th century. Some commentators place such work primarily in the realm of science fiction (Hamilton 2003; Clayton 2013), while a listing of works featuring biotechnology (self-admittedly still in an incomplete stage of construction) contains only a handful of examples (University of Bremen n.d.). But that is certainly changing, as documented in a recent book (Hamner 2017) that analyzes the influence of genetics in film and TV as well as fiction, including two of the novels[2] that had drawn my attention to the subject.

Indeed, the line between artificial/inorganic and biological creation of new forms of life has steadily blurred over the past few decades, with "bionic people" and "cyborgs" featuring in many movies, literary works, *etc.*, adding the category of "wetware" to those of software and hardware. One notable recent example amalgamates the story of Mary Shelley's conception and writing of *Frankenstein* with a contemporary quest for immortality by "uploading" the contents of a human brain into a machine (Winterson 2019); more on that later.

The 21st-Century Golem

In several of the afore-mentioned novels that I encountered in (mostly) recreational reading around the time I started paying attention to biotech in literature, golems unexpectedly appeared.[3] Two of those books seemed to have little if anything to do with biotech, except perhaps indirectly; but that *something* had provoked a renewed interest in golems was suggested by the frequency of their manifestation. The (only) book whose title did warn of golems to come, *The Golems of Gotham* (Rosenbaum 2002), has no obvious L&S connotations: its golems are animated in the traditional supernatural way, with incantations and numerology.

Much more surprising to me was to find a golem in Michael Chabon's *The Amazing Adventures of Kavalier and Clay* (2000). Its main subject is the story of two cousins, Josef Kavalier and Sammy Klayman, who play a key role in the comic book explosion following Superman. Again there is no science in the creation of life here. Indeed, the golem is not created in the book; it is never even animated. The premise is that the *original* golem of Prague had been preserved and hidden, and was sent abroad in 1939 to keep it from falling into the hands of the occupying Nazis; Josef is smuggled out of Prague in a coffin containing the golem. After that the golem appears in the novel only as a metaphor until the very end, when the coffin mysteriously turns up in Josef's house, containing just the loose dirt from which the golem had been made. But there *is* a brief allusion to its original animation, which—though traditionally supernatural—emphasizes a strong connection to language:

> Every universe, our own included, begins in conversation. Every golem in the history of the world, from Rabbi Hanina's delectable goat to the river-clay Frankenstein of Rabbi Jodah Loew ben Bezalel, was summoned into existence through language, through murmuring, recital, and kabbalistic chitchat—was, literally, talked into life.
>
> (119)

In contrast, the title character in Cynthia Ozick's *The Puttermesser Papers* (1997) actually does animate a golem. One evening, Ruth Puttermesser, totally dissatisfied with her life—her solitude and childlessness, her having been unfairly fired, her bleeding gums—finds a body in her bed "filmed with sand, or earth; some kind of clay" (38), with a piece of paper in its mouth bearing the Hebrew word for HaShem—"The Name" (standing for the name of God, which may not be spoken). When she reads it aloud, the creature comes to life, revealing itself to be a golem, even developing the classical trigrammaton אמת—in freckles—on its forehead (70). But this is a female golem, who chooses to be called Xanthippe (40–43), and who helps Puttermesser to great success—all the way to becoming Mayor of New York!—before going out of control and being destroyed by the traditional method of erasing the aleph (99).

Although this is again a supernatural creation, not scientific, there *is* an apparent allusion to biology, *via* yet another connection to language. Well before the appearance of Xanthippe, Puttermesser reflects:

> In bed she studied Hebrew grammar. The permutations of the triple-lettered root elated her: how was it possible that a whole language, hence a whole literature, a civilization even, should rest on the pure presence of three letters of the alphabet? The Hebrew verb, a stunning mechanism: three letters, whichever fated three, could command all possibility …. It seemed to her not so much a language for expression as a code for the world's design.
>
> (5)

Is it not likely that this is meant to draw attention to an analogy between language and the genetic code, likewise based on three-letter groups?

That analogy is expressed much more clearly in Harry Mulisch's *The Procedure* (2001). Here the main narrative concerns a contemporary biologist and his attempt to create life *de novo*. The author feels unable to plunge right into that story, starting the book with:

> Yes, of course I can come straight to the point …. But I can't do it that way this time. On the contrary. Before anything can come to life here, we must prepare ourselves through introspection and prayer.
>
> (3)

He then gives accounts—somewhat non-standard versions—of the creation of Adam in Genesis and of the golem of Prague. Like both Chabon and Ozick, he emphasizes the centrality of language in bringing new things to life, as well as highlighting the connection to science: words are to letters as molecules are to atoms:

> Listen. Of course you know that the worlds and Adam were created by the word … linguistic creation is not taken figuratively, as usually happens, but—with the inexorable consistency of Judaic mysticism—literally. Because words consist of letters, as molecules consist of atoms, we must focus attention on the elementary components' building blocks …. Twenty-two letters, he designed them, carved them out, weighed them, combined and transposed them, each with all; with them shaped the whole of creation and everything that remained to be created.
>
> (7–8)

In Mulisch's subsequent telling of the Prague golem story (19–53), Rabbi Löw creates the creature not on his own initiative, to protect the Jews, but rather to satisfy the curiosity of the emperor. After molding the being out of clay, he and his colleague Isaac set out to animate it by pronouncing all the syllables that can be made up by combining an initial *aleph* with all possible trigrammatons. They succeed—though not perfectly, as a mistake was made during the recitation:

A living being is lying there. The clay has become flesh .... On the forehead three letters have appeared, which he didn't put there: A M T ... " 'eMet,' " he reads. "Truth." ... Something's not right. Something's gone wrong. The creature has unmistakable female breasts ... the penis has not turned into flesh, but has become baked clay .... "Schlemiel!" With a sweep of his hand he hurls the terra-cotta penis into the Moldau. "Now I understand. After your stupid mistake with 'eL came 'aM, that is 'mother.' That obviously caused a short circuit. So we haven't imitated the creation of Adam, but that of Lilith."

(49)

The female golem (another one!) promptly kills Isaac and is in turn destroyed by Rabbi Löw in the traditional manner by removing the *aleph* from her forehead (51–52).

This seems another clear allusion to the connection between language and genetics: mutations are introduced by changing as little as a single letter in a recitation—or a single base in a DNA sequence. By extension, it also calls attention to the risks inherent in modification or creation of living species *via* biotechnology, which becomes a main topic when Mulisch turns to the story of Victor Werker. The choice of first name is surely no accident: Mulisch acknowledges his predecessor in a commentary on *Frankenstein* and how it came to be written (122–123).

In a series of letters to his daughter Aurora, Victor tells of his generation of living matter—which he calls the "eobiont"—out of microscopic crystals of clay. His account includes considerable background on the relevant science, including a table of the complete genetic code (114), and referring to many of the key scientists involved in the history of the field— Watson and Crick (86); Stanley Miller (87), whom we met in Chapter 9; Alexander Cairns-Smith (116), who hypothesized the intermediacy of clay crystals in the origin of life; and others—with fairly accurate capsule summaries of their contributions. He explains how changing one or more letters can lead to not only disastrous mutations—as with his golem—but also novel creations:

But a small mistake can be fatal .... That can result in your being born with a cleft palate, or later getting stomach cancer, like your grandfather. You'll now understand that by cutting and pasting those letters we are capable of the craziest things .... If we feel like it we can produce a waxworks of created fabulous creatures, which were previously reserved for fantasy: chimeras, basilisks, unicorns, griffins, centaurs, sphinxes, everything mankind had dreamed of.

(114–115)

The description of his actual achievement, however, is more *textual* than scientific:

> Let me try to give you an idea of our procedure with an example from typography. Let's say that ABC is inanimate clay and ABC is animate .... ABC consists of bare, sanserif letters with the same undifferentiated thickness everywhere, which is suitable for timetables, telephone books, and street signs, but not for literature .... You need serifs for that .... What we had to do was, therefore, by means of chemical cutting and pasting, to mix and stir that inorganic ABC, and by creating organic conditions change it into the organic ABC .... My little eobiont had seen the light of day: an extremely complex, chemically highly equipped organic clay crystal, with the character of proto-RNA, a sort of primeval ribosome, which produced a couple of short proteins so that my creature, provided with energy by sunlight, reproduced and had a metabolism.
>
> (118–119)

Indeed, throughout the book Mulisch juxtaposes scientific and literary creation, suggesting that they are similarly contingent and, importantly, *experimental*:

> I was convinced that a story could only appear to give a picture of a world in which free will and chance reign, because in a story even the smallest detail has been foreseen and is predestined by the omniscient, omnipotent, all-preordaining narrator .... But with hindsight the matter turns out to be not so simple .... The story itself is the actual narrator, it tells itself; from the first sentence onward the narrative is a surprise to the narrator too .... And in that sense, through its predictable/unpredictable manner of creation, through its free will, a story does after all give a faithful image of the world— just as physical reality itself is only determined to a certain extent by cause and effect, but in the last instant is subject to fundamental uncertainties, probabilities, chance.
>
> (13)

Having been invited to write a book about his discovery, Victor makes the parallel explicit:

> With a novelist ... I think that he writes his book the way that *book* wants to be written. It's a machine that builds itself and the reader has to make the best of it. Now the odd fact is that my eobiont is actually also a machine that builds itself, it's an organism, but I can't write in the same way about it, because with nonfiction that doesn't work.
>
> (110)

Later in the novel we learn that Aurora does not actually exist: she was strangled by the umbilical cord *in utero*. The titular "procedure" refers as much to the delivery of the stillborn body—which he is unable to watch (149–150)—as to his scientific accomplishment, about which he no longer much cares, recognizing the trade-off he has made after receiving an anonymous letter:

> Congratulations! Have you got what you wanted now? You paid for the eobiont with your own child. An eye for an eye, a tooth for a tooth! Soon it will be your own turn, you bastard!
>
> (156)

> Nietzsche's Zarathustra had a vision of God being dead—had he, Victor himself actually perhaps eliminated God a hundred years later by creating life? Is the soul of the commandment that thou shalt not kill perhaps the commandment that thou shalt not create life?
>
> (221)

At the end of the book he is stabbed to death—possibly deliberately, possibly the consequence of mistaken identity—as part of an explicitly Kafkaesque subplot. He welcomes it, with the word "dawn" alluding to the names of both his creation and his unborn daughter:

> On the second attempt the steel takes possession of his body and penetrates his heart .... After the mystery of life he has finally unveiled the mystery of death too .... Victor Werker is happy. The light of a dazzling dawn surrounds him. I am immortal, he thinks, as his eyes cloud over.
>
> (230)

The last few quoted passages all highlight a central theme of the book— the inability to achieve any semblance of control over one's life and world—that Mulisch finds equally applicable to biologists and novelists. The possibility—or impossibility—of control is a major topic of the books considered in the following section.

The (Re)Making of Mice and Men

We move on to the manipulation of more advanced organisms. Zadie Smith's debut novel *White Teeth* (2000) has as its main protagonists two families: those of Archie Jones, a native Brit, and Samad Iqbal, a Bengali immigrant, who have been friends since their shared experiences in WWII. The biotech theme is introduced with geneticist Marcus Chalfen, for whom both Archie's daughter Irie and Samad's elder (by two minutes) son Magid work as assistants. Chalfen is using genetic modification to generate "FutureMouse©," designed to develop tumors at pre-programmed times. This is the first stage in a program aimed ultimately at eliminating the role of chance in life, as he explains to Irie:

> But if you *re-engineer* the actual genome, so that *specific* cancers are expressed in *specific* tissues at *predetermined* times, then you're no longer dealing with the *random* …. Now you're talking the genetic program *of the mouse* …. You eliminate the random, you rule the world …. One could program every step in the development of an organism: reproduction, food habits, life expectancy …. WORLD DOM-IN-A-SHUN …. Are you following me? One mouse sacrificed for 5.3 billion humans. Hardly mouse apocalypse. Not too much to ask.
>
> (340–341)

a project to which Magid signs on enthusiastically:

> Magid was proud to say he witnessed every stage. He witnessed the custom design of the genes. He witnessed the germ injection. He witnessed the artificial insemination. And he witnessed the birth, so different from his own. One mouse only. No battle down the birth canal, no first and second, no saved and unsaved. No potluck. No random factors. No *you have your father's snout and your mother's love of cheese*. No mysteries lying in wait …. No second-guessing, no what-ifs, no might-have-beens. Just certainty. Just certainty in its purest form. And what more, thought Magid—once the witnessing was over, once the mask and gloves were removed, once the white coat was returned to its hook—what more is God than *that*?
>
> (489–490)

But Smith's world—like Mulisch's—emphatically does *not* work that way. There can be no certainty, no elimination of random factors. We see this from the very opening scene, in which Archie contemplates killing himself. When a coin flip comes out in favor, he tries to gas himself in his car; but he happens to have parked it in a butcher shop's delivery zone, and the owner saves him:

> Mo advanced upon Archie's car, pulled out the towels that were sealing the gap in the driver's window, and pushed it down five inches with brute, bullish force. "Do you hear that, mister? We're not licensed for suicides around here. This place halal. Kosher, understand? If you're going to die round here, my friend, I'm afraid you've got to be thoroughly bled first."
>
> (7)

Chalfen's career itself depends largely on accident: he works in the Perret Institute, led by Dr. Perret, who (we learn) had done medical research for

the Nazis during WWII before being captured by Russians. They hand
him over to Samad, who tells Archie to kill the man he calls "Dr. Sick":

> Do you know who this man is, Jones? .... He's a scientist, like me—
> but what is his science? .... He wants to control, to dictate the future.
> He wants a race of men, a race of indestructible men, that will sur-
> vive the last days of this earth.
>
> (119)

But Archie decides to spare him by (how else?) flipping a coin (539–540),
allowing Chalfen's eventual mentor to survive by chance, although he
tells Samad he *did* carry out the execution.

As the plot proceeds, two organizations that are opposed to Chalfen's
project appear. One is a Muslim fundamentalist group led by Magid's twin
brother, Millat, who consider such playing with life to be blasphemous—
as made explicit in Magid's reference to God in the quote on p. 172. Even
the name chosen by the group demonstrates how hard it is to keep control over one's world:

> "He's the head of the Cricklewood branch of the Keepers of the
> Eternal and Victorious Islamic Nation." The headmaster frowned.
> "KEVIN?" "They are aware they have an acronym problem,"
> explained Irie.
>
> (301)

Chalfen's son Joshua becomes involved with the other group, Fighting
Animal Torture and Exploitation (a more suitable acronym). All the
factions converge, towards the end of the book, at the unveiling of
FutureMouse©. On the way there, Joshua reflects on his philosophical
differences with his father:

> What he was about to do to his father was so huge, so *colossal*, that
> the consequences were inconceivable .... Something like the end of
> the world .... Joshua glared up and down Whitehall, at the happy
> people going about their dress rehearsal. They were all confident that
> it wouldn't happen or certain they could deal with it if it did. But
> the world happens to you, thought Joshua, you don't happen to the
> world. There's nothing you can do. For the first time in his life, he
> truly believed that. And Marcus Chalfen believed the direct opposite.
>
> (497)

At the event Perret is introduced, revealing to Samad that Archie had
lied to him all those years ago. Millat moves forward to shoot Perret,
but manages to shoot Archie instead, and the book ends with an entirely
unexpected (except, possibly, by the mouse itself) outcome: the escape of
the one and only FutureMouse©.

So Archie is there, there in the trajectory of the bullet, about to do something unusual, even for TV: save the same man twice and with no more reason or rhyme than the first time .... Everybody in the room watches in horror as he takes it in the thigh, right in the femur, spins around with some melodrama and falls right through the mouse's glass box .... It would make an interesting survey (what kind would be your decision) to examine the present and divide the onlookers into two groups: those whose eyes fell upon a bleeding man, slumped across a table, and those who watched the getaway of a small brown rebel mouse. Archie, for one, watched the mouse. He watched it stand very still for a second with a smug look as if it expected nothing less. He watched it scurry away, over his hand. He watched it dash along the table, and through the hands of those who wished to pin it down. He watched it leap off the end and disappear through an air vent. *Go on my son!* thought Archie.

(540–542)

Despite the fact that *White Teeth* is a thoroughly comic novel (as can be easily seen from the irresistible quotes), the pitfalls of biotech are represented and integrated with the other main themes in a remarkably coherent and effective manner.

Lastly, we have two books that take us all the rest of the way up the evolutionary ladder, to *human* life. Their central premise is much the same. In each, a brilliant biotechnologist decides that society must not continue as it is, and brings about the replacement of humanity with a posthuman successor, although the particular societies that are undone, and the means of carrying it out, differ considerably. The two works reveal interesting parallels as well as contrasts, both with those discussed earlier and with each other.

In Michel Houellebecq's *The Elementary Particles* (2000) the unacceptable society is our own current one. There are two main characters—half-brothers Bruno Clément and Michel Djerzinski—who both illustrate the author's apparent view of the contemporary human condition: disaffection; isolation; the impossibility of meaningful relationships with other humans. Michel is a biotechnologist, initially engaged in commercial activities. Bruno tells someone that Michel's "research led to the development of genetically modified cows which produce more milk .... He changed the world" (168). That last phrase is just Bruno's self-deprecatory comparison with his own lack of accomplishments; but in fact Michel *does* go on to totally change the world. His loathing for what he calls the "atomization of society" (129; an earlier publication of the same translation in the UK was titled "Atomised") leads him to quit his job and go off to several years of solitary thinking and research, aimed at what he sees as the cause of problem: the unreliability of the reproductive mechanism. He generates a series of groundbreaking papers, culminating in "Toward a Science of Perfect Reproduction" (136), after which he (perhaps; it is left somewhat ambiguous) kills himself (253).

In an epilog, set around 2050, we are informed that Michel's disciple, Hubczejak, formed a movement that implemented those theories to create a new race of posthumans—asexual, genetically identical, and immortal—that essentially replaced existing humanity. The (unidentified) author of the epilog considers this a positive, even essential development:

> [E]very member of the species created by making use of Dzerjinski's work would carry the same genetic code, meaning that one of the fundamental elements of human individuality would disappear. To this Hubczajek responded that this unique genetic code—of which, by some tragic perversity, we were so ridiculously proud—was precisely the source of so much human unhappiness.
>
> (261)

> [A] fundamental shift was indispensable if society was to survive—a shift which would credibly restore a sense of community, of permanence and the sacred .... they believed in their hearts that the solution to every problem—whether psychological, sociological, or more broadly human—could only be a technical solution.
>
> (262)

> Today, some fifty years later, reality has largely confirmed the prophetic tone of Hubczajek's speech .... There remain some humans of the old species .... This vile, unhappy race, barely different from the apes, which nevertheless carried within it such noble aspirations. Tortured, contradictory, individualistic, quarrelsome and infinitely selfish .... This species which, for the first time in history, was able to envision the possibility of its succession and, some years later, proved capable of bringing it about.
>
> (263–264)

It is not clear to what extent, if at all, such statements represent Houellebecq's view. Is it significant that the author gave his own first name to his biotechnologist? His other character Bruno endorses the obvious similarities between Houellebecq's vision of the future and Huxley's in *Brave New World*:

> Everyone says *Brave New World* is supposed to be a totalitarian nightmare, a vicious indictment of society, but that's hypocritical bullshit. *Brave New World* is our idea of heaven: genetic manipulation, sexual liberation, the war against aging, the leisure society. This is precisely the world that we have tried—and so far failed—to create.
>
> (131)

It is hard to see Bruno (or Michel, for that matter) as anything other than thoroughly despicable, so perhaps we should take this comment as an exhortation *not* to think that way. But who knows? A reviewer commented about a later novel: "This writer's characters are, in their way, moral beacons for our era. In nearly any situation, one can ask 'What would a Houellebecq man do?' and perform the opposite" (Garner 2019).

The second book, Margaret Atwood's *Oryx and Crake* (2003), is set sometime later in the 21st century, in a world that is an exaggerated version of our own, but (unfortunately) not unrecognizably so. Technology—especially biotech—and capitalism have run rampant, to the extent that when the bio/pharma/health care industry has been so successful at finding cures for diseases that its own income becomes threatened, it develops and spreads new ones (247–248). Crimes meriting execution include "hampering the dissemination of commercial products" (337). The country has been divided into "Compounds"— gated and secured areas where all the important, productive people live and work—and "pleeblands," home to everyone else. New life forms are everywhere: "rakunks" (raccoon/skunk crosses to make superior pets), "pigoons" (pigs modified for enhanced growth of human body parts for replacements), and many others.

The narrative is presented in two alternating streams—before and after the apocalypse—from the point of view of Jimmy (later "Snowman"). He starts as a privileged resident of the Compounds because his parents are successful biotechnologists, although he himself has little or no aptitude for science and math, being much more language-oriented. His classmate Crake,[4] in contrast, has outstanding abilities in the important fields, and moves on to the Watson-Crick Institute (whereas Jimmy can't do better than a much inferior university for artsy types, the Martha Graham Academy) and a job at the highly prestigious biotech firm RejoovenEsense. There he has developed a new drug called BlyssPluss that vastly increases sexual activity (346–347). But his main interest—the Paradice Project— is the creation of a new race of humans, designed to eliminate all the problems of human history:

> What had been altered was nothing less than the ancient primate brain. Gone were its destructive features, the features responsible for the world's current illnesses. For instance racism ... the Paradice people simply did not register skin color. Hierarchy could not exist among them, because they lacked the neural complexes that would have created it .... They ate nothing but leaves and grass and roots and a berry or two; thus their foods were plentiful and always available. Their sexuality was not a constant torment to them, not a cloud of turbulent hormones: they came into heat at regular intervals, as did most mammals other than man.
>
> (358–359)

Both of Crake's projects are ostensibly commercial ventures—he brings Jimmy into the company supposedly to write advertising copy for the former—but in actuality his goal is to replace humanity with the new "Children of Crake." He has also developed a new, universally fatal virus, which he surreptitiously incorporates into the BlyssPluss pills that are distributed throughout the world, killing off *everyone*[5]—except Jimmy, whom he has secretly treated with a vaccine, so that he can survive and help shepherd his new creations as they inherit the earth.

The representation of science in both of these works is considerably more elaborate than in those discussed previously. Atwood puts in considerable effort to portray most of the developments as reasonable extrapolations—or even actual examples—of contemporary biotech. For example, a species of rabbit that glows in the dark escapes into the wild and becomes a nuisance (109–110). Such a transgenic animal (named "Alba") had already been created by the time Atwood wrote—a joint project of an artist and geneticists—by making use of a jellyfish-derived gene (green fluorescent protein, or GFP) much used in biological research (Young 2000). She also anticipates the burgeoning DIY or biohacking movement mentioned earlier: an informal association that calls itself MaddAddam carries out biosabotage on a major scale—"the asphalt-eating microbes, the outbreak of neon-coloured herpes simplex"—and is recruited by Crake for his projects (352). On the other hand, the possibility of making specific programmed changes to the brain circuitry as described in the quote on p. 176 goes *far* beyond anything currently imaginable. But as both our genetic toolkit and our understanding of brain function continue to expand, it is perhaps not such an inconceivable development.

Houellebecq also extensively incorporates scientific concepts into his narrative, but little of it is based on real, relevant science. Michel's central idea is taken more from topology than biotech: he concludes that the double helix structure of DNA is inherently unstable, and needs to be replaced by genetic material based on a different structure:

> [He] established, for the first time, on the basis of irrefutable thermodynamic arguments, that the chromosomal separation at the moment of meiosis which creates haploid gametes is in itself a source of structural instability. In other words, all species dependent on sexual reproduction are by definition mortal.
>
> (248)

This, like nearly all of Houellebecq's purported scientific expositions, is total nonsense. He tries to ground his "theories" by citing real scientists and philosophers—Einstein, Cantor, Comte, *etc.*—but most of that exposition is incoherent and/or irrelevant. Just to cite one example: Michel takes from a mentor the idea that quantum theory will be crucial to biotech advances:

> Desplechin pointed out that it was the existence of a finite list of chemical elements which had prompted Niels Bohr's first thoughts on the subject .... It was this situation, profoundly anomalous according to Maxwell's equations and the classical laws of electromagnetics, that led to the development of quantum mechanics. In Desplechin's opinion, contemporary biology was now in a similar position. The existence of identical macromolecules and immutable cellular ultrastructures, which had remained consistent throughout evolutionary history, could not be explained by the laws of classical chemistry. In some way as yet impossible to determine, quantum theory must directly impact on biological events.
>
> (104)

This is incorrect in every possible way. Bohr's work was inspired not by the finite number of elements, but by the fact that according to existing (pre-quantum) theory atoms should not be stable. (Ironically, that would have been a much better analogy for the idea that the instability of DNA during reproduction is the problem that needs fixing!) There are *no* "laws of classical chemistry" separable from quantum theory, and in that sense the impact of the latter on biology is well understood. Any number of similarly confused and pretentious passages could be cited. Unlike Atwood, Houellebecq makes no serious attempt to come to terms with the latest science and its consequences for the human condition; he merely tosses out a potpourri of buzzwords in the hope of sounding relevant and up to date.[6]

The treatment of the other two common themes, control and language, also differs considerably between these two end-of-humanity novels and the rest. The golem tales and (especially) *White Teeth* emphasize the inability to control any desired outcome: chance, uncertainty, and unintended consequences rule. Both Michel and Crake aim for elimination of random factors; but unlike Chalfen, Victor Werker, and the other golem creators, they appear to succeed, to a substantial degree. As we saw earlier, a stable posthuman society *does* result from Michel's work, 50 years later. *Oryx and Crake* describes many examples of unintended consequences of the biotech done by Crake's predecessors; but Crake's creatures apparently come out pretty much as planned. At one point Jimmy (as Snowman), observing the "Crakers" after everyone else has died, comments on how Crake seems to have anticipated every contingency that they are likely to encounter, even relatively minor ones:

> Sitting all together like this, they smell like a crateful of citrus fruit—an added feature on the part of Crake, who'd thought those chemicals would ward off mosquitoes. Maybe he was right, because all the mosquitoes for miles around appear to be biting Snowman.
>
> (117)

But has he really? Complete control is surely not what Atwood means to convey. The very name of the project "Paradice" suggests considerable ambiguity on that score, and a scene towards the end hints at contravening Crake's intentions:

> They're sitting in a semi-circle around a grotesque-looking figure .... *Ohhhh*, croon the women. *Mun*, the men intone. Is that *Amen*? Surely not! Not after Crake's precautions, his insistence on keeping these people pure, free of all contamination of that kind. And they certainly didn't get that word from Snowman. It can't have happened.
> (418–419)

Indeed, if any idea of taming the inherent impossibility of control through biotechnology *is* to be found in *Oryx and Crake*—as it is, more clearly, in *The Elementary Particles*—it is totally subverted in two sequels published subsequently (2009, 2013).[7]

I find the most striking difference between the two last-discussed books and the others—also the most interesting from the L&S point of view—in the relationship between science and words. We saw in the golem books a strong *connection* between the two, deriving from the central role of words (spoken and/or written) in the ritual of animation, and the analogy to the genetic code drawn by some of the authors. In that sense language and science are portrayed as working together. In contrast, both Houellebecq's and Atwood's future worlds represent the *dis*connection of science from words—more precisely, the triumph of the former over the latter. Houellebecq makes that explicit, while casting not-at-all subtle aspersions on some of his French intellectual competitors:

> The global ridicule in which the works of Foucault, Lacan, Derrida and Deleuze has suddenly foundered, after decades of inane reverence, far from leaving the field clear for new ideas, simply heaped contempt on all those intellectuals active in the "human sciences." The rise to dominance of scientists in all fields of thought became inevitable.
> (262)

In *Oryx and Crake* the immense gap between the prospects available to Crake, a scientist, and Jimmy, a wordsmith, demonstrates the hegemony of science. (The infantilized spellings, some of which have been seen earlier, may also be intended to depict the degradation of language.) Crake intentionally deprives his creatures of any written language, and limits their verbal abilities; after Jimmy/Snowman no longer has anyone to talk to, the demise of language will be complete:

> "Hang on to the words," he tells himself. The odd words, the old words, the rare ones. *Valence. Norn. Serendipity. Pibroch.*

Lubricious. When they're gone out of his head, these words, they'll be gone, everywhere, forever. As if they had never been.

(78)

Frankisstein also endorses the power of language: the protagonist, a cyberscientist, quotes Camus, "To name things wrongly is to add to the misfortunes of the world," and continues: "I can tell you that calling things by their right names is more than giving them an identity bracelet or a label, or a serial number. We summon a vision. Naming is power" (Winterson 2019, 79). In an afterword/interview, the author suggests that a significant component of Victor Frankenstein's reprehensible treatment of his creature was failing to give it a name (354).[8]

On this score, then, these books point in opposite directions. Two—Atwood and Houellebecq—fear (or perhaps, in the latter case, hope?) that biotechnology heralds the decline and ultimate demise of language: as we move towards the posthuman, we will also move post-literature. In contrast, by highlighting the perpetuation of earlier traditions, the other authors in effect connect the creation of life with the creation of literature, suggesting that ongoing developments in biotechnology—the "new, new thing"—will be matched by parallel evolution of L&S. Hopefully they will prove to be more prescient.

Notes

1 This chapter is based on a talk presented at the 2004 (Paris) SLS-Euro meeting.
2 Hamner devotes a good deal of attention to two of these (*White Teeth* and *Oryx and Crake*), although his main focus differs somewhat from mine.
3 I may have been sensitized to connections between the golem and L&S by the book cited in Chapter 3, which was one of my early encounters with science studies. The authors suggested, more than a little fancifully, that the golem legend could serve as a useful analogy for the workings of science:

> [A golem] is powerful .... It will follow orders, do your work, and protect you from the ever threatening enemy. But it is clumsy and dangerous ... it is truth that drives it on. But this does not mean it understands the truth—far from it ... it is not an evil creature but it is a little daft. Golem Science is not to be blamed for its mistakes; they are our mistakes. A golem cannot be blamed if it is doing its best. But we must not expect too much.
>
> (Collins and Pinch 1993, 1–2)

I thought at the time (and still do) that was a singularly inapt metaphor (Labinger 1993), but I will not go into that here.
4 Not his real name, of course, but one he takes on as a player of a computer game called Exinctathon, whose aficionados name themselves for threatened or vanished animals. His parents named him Glenn, after Glenn Gould (80). Is this just Atwood's bow to a fellow Canadian? No: Atwood saw parallels between Crake and Gould (Elliot 2006, 824), just as Powers did between Ressler and Gould in *The Gold Bug Variations* (see Chapter 5).

5 Including Crake himself: he (almost certainly intentionally) induces Jimmy to shoot him by killing their mutual love interest Oryx, presumably not wanting to live on to see his new world. Yet another parallel with Houellebecq's Michel.

6 I may have given the impression that I did not like this book; that would be entirely accurate. In nearly seven decades of reading I can recall no other book that I disliked so much and still stuck with all the way to the end. I did so because of the relevant biotech themes under consideration here, and have done my best to address those impartially; but in full disclosure I should acknowledge my opinion (which others share: some typical reviewer characterizations include "a deeply repugnant read" (Kakutani 2000), "bilious, hysterical and oddly juvenile" (Quinn 2000), and so on).

7 I will not address those here; a useful discussion of how these issues develop over the course of the trilogy is offered in Hamner (2017, 147–173).

8 Winterson in contrast names her characters pointedly, if not very subtly (subtlety was clearly not one of her aims in this book): the main contemporary characters are Ry (short for Mary: s/he is transgender) Shelley, Victor Stein, Ron Lord, Claire, and Polly D. All these refer to the house-party at Byron's villa in Switzerland where Mary Shelley began to write *Frankenstein* in 1816: her companions (besides her husband-to-be Percy) were Lord Byron, her step-sister (and Byron's mistress) Claire, and Byron's physician Dr. Polidori.

11 The End of Irony and/or the End of Science?[1]

This chapter has had several sources of inspiration over a number of years. The first was a book that I read shortly after I had gotten seriously interested in L&S. *The End of Science* (written by a science writer, not a practicing scientist) proclaimed that "the great era of scientific discovery is over" (Horgan 1996, 6). At the time the book didn't strike me as fruitful material for my own L&S work, although Horgan did draw some connections between science and literary criticism, noting that he began as an English major who thought of the latter as "the most thrilling of intellectual endeavors" (3). I contented myself with writing a review (quite negative, for reasons that will be seen shortly) for a Caltech journal (Labinger 1996b).

The next stimulus, which came a few years later, was the proclamation of an "end to irony" in reaction to 9/11. Here is a typical example:

> One good thing could come from this horror: it could spell the end of the age of irony. For some 30 years—roughly as long as the Twin Towers were upright—the good folks in charge of America's intellectual life have insisted that nothing was to be believed in or taken seriously. Nothing was real. With a giggle and a smirk, our chattering classes—our columnists and pop culture makers—declared that detachment and personal whimsy were the necessary tools for an oh-so-cool life. Who but a slobbering bumpkin would think, "I feel your pain"?
>
> (Rosenblatt 2001)

This new prognostication of an "end" reminded me of Horgan. Just as I had been highly skeptical of Horgan's forecast for the future of science, it seemed to me that irony had survived events at least as earthshaking as 9/11, and was quite likely to do so again. Others felt the same:

> In the anguished days following Sep. 11 cultural prognosticators were quick to declare the death of irony, cynicism and black humor .... Many of the forecasts mistook shock and grief for long-term cultural change and have already been proven wrong ... the belief that

DOI: 10.4324/9781003197188-11

the terrorist attacks will lead to kinder, gentler entertainment belies the historical record of reactions to earlier tragedies, wars and social upheavals.

(Kakutani 2001)

Indeed, fictional works dealing with 9/11 began to appear within a few years (Wyatt 2005)—some of them, no doubt, at least somewhat ironic—while a number of pundits proposed the replacement of the "age of irony" with a "New Sincerity" movement, which had begun well before 9/11, and would "overturn the ironic detachment that hollowed out contemporary fiction towards the end of the 20th century" (Moats 2012).

Reference to these "ends"—of science and of irony—established only a relatively small degree of resonance; but I spotted a stronger one. Horgan had coined the phrase "ironic science" as an important component of his thesis, and I wondered whether the common appearance of "irony" in these doomsayings was significant, or just a coincidental choice of wording. However, I was not moved to any serious exploration until a couple of years later, when in a short period of time I happened to read three novels—*Prague* (Phillips 2002), *The Double* (Saramago 2004), and *Everything Is Illuminated* (Foer 2002)—that were deeply saturated in ironic tone. Although all three had certainly been written mostly or entirely before 9/11, the experience (accidental though it may have been) reminded me of my thoughts about joining up the ends of irony and science, and started me contemplating whether these books might be connectable as well.

In this chapter I will first present Horgan's thesis, including an attempt to unpack just what he means by "ironic science," and explain why I find it completely unconvincing. I will then briefly examine the generally accepted meaning(s) of irony, and how they might relate to Horgan's concept. Finally, I will discuss the role of irony in the three books, and offer an interpretation for fitting together all these "ends" and "ironies."

The End of Science?

Horgan, formerly a science writer for *Scientific American*, is convinced that the golden age of science is coming to an end. Why? He offers three main reasons. One is rather mundane: science is running up against the law of diminishing returns. Experiments are becoming harder and (especially) more expensive, at the same time that society is becoming less willing and/or less able to fund them. Certainly there has been evidence of such a trend, starting well before 1996 and continuing to the present day; but is it both universal and irreversible? Some very large-scale and expensive programs *have* been funded (LIGO, the search for gravity waves mentioned in Chapter 3, is a prime example) even while others have found themselves under severe economic pressure. I don't see how one can assert with confidence that groundbreaking science will not be carried out for lack of resources.

Horgan's second reason is really a value judgment: in his view, all the big problems have been solved, or soon will be. There just aren't many more truly fundamental discoveries—like quantum mechanics, relativity, evolution, the big bang, DNA—left to be made, so only less interesting activities will remain, such as exploring the detailed implications of the basic theories, working on applied problems, and so on. Not that he considers them *unworthy*, exactly; but he feels they don't truly fire the imagination and attract the kind of intellectual superstars responsible for the triumph of science over the last couple of centuries.

That was not a new idea: around the end of the 19th century—shortly *before* all the great discoveries Horgan enumerates were made—it was commonly thought that there was little left to do beyond adding the next digits after the decimal point. Horgan's characterization of contemporary science is strikingly similar: "Of course, science will continue to raise new questions. Most are trivial, in that they concern details that do not affect our basic understanding of nature. Who really cares, apart from specialists, about the precise mass of the top quark?" (30). Still, he defends himself against the charge of repeating history by noting that a) not *everybody* believed it back then, and b) even if it was wrong then, it doesn't follow that it is wrong *now*. I suppose that is logically correct, but it is not very satisfying. More importantly, though, his rankings of scientific discoveries—great *vs.* just so-so—seem quite arbitrary. A typical example: "Quantum mechanics ... was an enormous surprise .... The later finding that protons and neutrons are made of smaller particles called quarks was a much lesser surprise, because it merely extended quantum theory to a deeper domain" (17). *Merely*??

To be fair, Horgan does not offer these value judgments as his alone: he supports his stance with interviews of some 45 prominent scientists, drawn from a variety of fields—philosophy, physics, cosmology, evolutionary biology, social science, neuroscience, "chaoplexity," "limitology," and machine science. (I was initially encouraged to see chemistry omitted from the list, but further reading dispelled that thought: according to Horgan, chemistry had *already* reached its end "in the 1930s, when the chemist Linus Pauling showed how all chemical interactions could be understood in terms of quantum mechanics" (10), so it was not even worth talking about any more! It might also be noted, though, that in a subsequent article (Horgan 2004, 40) he acknowledged that he had "always found chemistry excruciatingly dull.") He announces that most (but by no means all) of his interlocutors concur that the glory days of science—at least, of their own branch of science—are coming to an end.

But Horgan's reportage is distorted in two significant ways. First, I was struck by the fact that the vast majority of Horgan's interview subjects had reached or were nearing the end of their *own* scientific careers. Their average age at the time of interview was about 65, and around half a dozen had died by the time the book appeared. Perhaps it is not too cynical to suspect that being on the verge of exiting a field predisposes one to lament the passing of its heyday? Also, Horgan makes not even a pretense of impartial presentation. On the (rare) occasion when he does give space to an opposing point of view, he manages to do so disparagingly. For example: "In denying the implication of his own ideas [that science might be ending], Chomsky may have been exhibiting just another odd spasm of self-defiance" (153). Elsewhere, Horgan applies the term "patronizing" to Thomas Kuhn's description of normal science as puzzle-solving (7), when not 40 pages later he quotes Kuhn as explicitly denying any such intent (45). Physicist Leo Kadanoff is quoted approvingly in support of Horgan: "Studying the consequences of fundamental laws is 'in a way less interesting' and 'less deep' ... than showing that the world is lawful" (27), while Stephen Jay Gould's contrary suggestion that "[fundamental] laws do not have much explanatory power; they leave many questions unanswered" is dismissed as "ironic science in its negative capability mode" (126).

That brings us to Horgan's third argument for the End of Science—the one most relevant to this chapter—which introduces the concept of ironic science. Although *he* is convinced that only the less interesting activities, such as exploring the detailed implications of the basic theories, working on applied problems, and so on, remain to be done, he recognizes that there are still some who believe they are working on big problems. Nonetheless, he claims, they too are encountering the intrinsic limits of science, because they are asking only questions that they will never be able to answer:

> Optimists who think they can overcome all these limits ... have only one option: to pursue science in a speculative, postempirical mode that I call ironic science. Ironic science resembles literary criticism in that it offers points of view, opinions, which are, at best, interesting, which provoke further comment. But it does not converge upon the truth. It cannot achieve empirically verifiable surprises that force scientists to make substantial revisions in their basic description of reality.
>
> (129)

Clearly Gould's statement, which Horgan brushed off so casually, does not at all fit this definition. It is neither speculative nor "postempirical," but rather, a straightforward opinion, which should merit as much respect as anyone's. But even leaving that out, this third argument of Horgan's is no more convincing than the others.

It is worth noting that a second edition of *The End of Science* appeared in 2015. In the Preface thereto Horgan basically doubles down on his position, claiming that none of his predictions had been disproven, and dismissing any new advances. For example, the discovery that the rate of expansion of the universe is accelerating—which he admits is an "astonishing finding"—is presented as just another "twist" of understood cosmology (*xvii–xviii*). He summarizes his stance in a statement that I find remarkably self-contradictory: "[N]othing can shake my faith in my meta-argument: Science gets things right, it converges on the truth, but it will never give us *absolute* truth" (*xxiv*). That seems to me a rather convincing argument that there will *never* be an End of Science.

Irony and Self-Reference

Why *did* Horgan hit upon "ironic" as a descriptor for the contemporary science he disparages, and how—if at all—could the word be applicable to all these cases: unverifiable science, reactions to 9/11, and the novels that were the ultimate inspiration for this chapter? To be sure, "irony" is a highly polyvalent term; many commentators have analyzed its meanings and usages at considerable length. One starts off in ironic mode, thereby self-exemplifying a readily recognizable form of irony:

> It might perhaps be prudent not to attempt any formal definitions. Since, however, Erich Heller, in his *Ironic German*, has already quite adequately not defined irony, there would be little point in not defining it all over again.
>
> (Muecke 1969, 14)

Another points to the difficulty of *any* definition:

> Despite its unwieldy complexity, irony has a frequent and common definition: saying what is contrary to what is meant ... our very historical context is ironic because today nothing really means what it says. We live in a world of quotation, pastiche, simulation and cynicism: a general and all-encompassing irony. Irony, then, by the very simplicity of its definition becomes curiously undefinable.
>
> (Colebrook 2004, 1)

Nonetheless, both proceed—at full book length—to enumerate classifications and illustrations. The Colebrook quote captures at least some of what I take the 9/11 responses to refer to—an amused, superior, detached worldview that resists taking anything too seriously—but that hardly seems relevant to Horgan's concept. Indeed, Muecke observes that some analysts have considered *all science* to be ironic, which surely Horgan would not accept either:

> Like the scientist, the General Ironist has a need and a capacity for endless revision and self-correction, for questioning and suspending judgment, for living "hypothetically and subjectively" as Kierkegaard says, and keeping alive a sense of an infinity of possibilities. According

to Goethe one must have a sense of irony to be a good scientist, since any results one arrives at may be superseded.

(129)

Can we find a definition or interpretation of irony that *does* seem consistent with "ironic science?" Horgan gives us a clue, in explaining his transformation from an English major to a science writer:

> [O]ne of the messages of modern criticism, and of modern literature was that all texts are "ironic": they have multiple meanings, none of them definitive .... Arguments over meaning can never be resolved, since the only true meaning of a text is the text itself.
>
> (1996, 3)

The last phrase invokes *self-reference* as a central focus of irony; in other words, Horgan seems to be suggesting that ironic science is about itself instead of about the real world. There are many passages in his book that support this interpretation. String theory (Greene 1999) is perhaps his prime exemplar:

> Physicists working on superstring theory, Lindley contended, were no longer doing physics because their theories could never be validated by experiments, but only by subjective criteria, such as elegance and beauty .... In what sense will such a theory explain the world? .... [A superstring] is some kind of mathematical ur-stuff that ... does not itself correspond to anything in our world.
>
> (70–71)

but there are others as well:

> There is no better example of a sophisticated ironic scientist than the anthropologist Clifford Geertz ... his work is one long comment on itself.
>
> (154)

> C]haoplexologists have created some potent metaphors .... But they have not told us anything about the world .... Computer simulations represent a kind of metareality ... but they are not reality itself.
>
> (226)

Indeed, self-reference has been identified as a common feature in more general treatments of irony. Muecke identifies it as the hallmark of what he terms Romantic Irony:

> There is potential for irony in the very nature of art if we regard it ... as both a communication and the thing communicated, that is,

as meaningful in its relation to the ordinary world and also as pure meaningless existence in itself .... The sophisticated or self-conscious artist ... will sometimes bring into his work at the imaginative level some aspects of its existence ... as a work that is being composed .... That is to say he will break into the artistic illusion with a reminder to his public (not necessarily an explicit one) that what they have before them is only a painting, a play, or a novel and not the reality it purports to be. This sort of thing has been called Romantic Irony.

(163–164)

More recently it has been associated with our postmodern age and the "end of irony" diatribes:

Our age has not so much redefined irony, as focused on just one of its aspects. Irony has been manipulated to echo postmodernism. The postmodern, in art, architecture, literature, film, all that, is exclusively self-referential—its core implication is that art is used up, so it constantly recycles and quotes itself. Its entirely self-conscious stance precludes sincerity, sentiment, emoting of any kind, and thus has to rule out the existence of ultimate truth or moral certainty.

(Williams 2003)

Both Horgan and Williams here give self-reference a strongly negative connotation; is that merited? Other literary critics have described very *positive* functions; one is to strengthen our ability to "compartmentalize," to carry out multiple mental functions at the same time:

[R]eflexive fictions ... give an interesting workout to our capacity for simultaneous trust and distrust, readying us for the difficult business of life. Yes, that difficult business often involves knowing the truth. But at times it requires us to be ignorant or even frankly mistaken—and when that happens, it is generally better for us to maintain an awareness of what is going on.

(Landy 2015, 566)

Mathematics gives us a well-known example where self-reference is explicitly involved. The proof of Gödel's Incompleteness Theorem—that any consistent formal system of arithmetic will include propositions that are true but unprovable—relies on a numbering scheme that causes statements about arithmetic also to be statements about themselves. Horgan refers to Gödel's theorem repeatedly (no fewer than 16 citations in his index!), offering it in support of his thesis that "science itself, as it advances, keeps imposing limits on its own power" (5). He apparently feels that this demonstration of a truth that is achieved through self-reference undermines the possibility of reaching ultimate, complete understanding. But as

I commented earlier, that doesn't imply any sort of *end* of science; on the contrary, it is an argument that science *can't* reach an end.

Indeed, irony has been associated with open-endedness: "Irony works against its own striving or intention for completeness, aware that such a striving can only fail, but that the failure is itself a moment of partial illumination" (Colebrook 2004, 68). And (ironically, as Rebecca Goldstein points out) Gödel himself would surely not have been on Horgan's side:

> One of the strange things that happened in the twentieth century was that results from mathematics and physics got co-opted into the assault on objectivity and rationality. I'm thinking primarily of relativity and Gödel's incompleteness theorems ... the irony is that both Einstein and Gödel ... could not have been more committed to the idea of objective truth .... The irony is sharpened in Gödel's case since not only was he a mathematical realist, believing that mathematical truth is grounded in reality, but, even more ironically, it was this meta-mathematical conviction that actually motivated his famous proofs.
>
> (Goldstein 2005)

Moving from the realm of mathematics to that of self-referential "real" science, we need to pose a key question: what is the "self" that "self-reference" refers to? Horgan (and/or his interviewees) characterizes string theory as a mathematical abstraction disconnected from the real world; but its implications must at least be *consistent* with what physicists already know about the world, even if those points of consistency are as yet far too limited to verify the theory. Just as "No man is an island," no theory is a completely isolated construct. Perhaps a better reference point than Donne's is a quote from Ortega y Gasset, "Yo soy yo y mi circunstancia" ([1914] 2004, 757)—I am myself *and* my circumstances. The self is inextricably coupled to its environment; string theory is about itself *and* its circumstances—the latter being the known physical world.

In that sense, how is string theory, or any of Horgan's "ironic sciences," different from those he lauds as the great theoretical discoveries of the past—relativity, quantum mechanics, *etc.*? In Horgan's view, the latter could be (and were) experimentally confirmed and thus connected to the real world, whereas that has not yet been and (he believes) never will be possible for string theory. That prediction seems remarkably pessimistic to me (admittedly, I am no physicist). A theory may not be empirically verifiable today, but who knows what new techniques and methodology might become available in the future? Indeed, many components of early 20th-century quantum mechanics could well have been characterized as ironic science by Horgan's criteria at the time they were introduced, as there were then no experiments available for their validation. Today they represent prime exemplars of the science to which Horgan pays his greatest homage.

Going further back in science history—of chemistry in particular—molecular structural theory, examples of which we have seen in preceding chapters, was introduced in the 19th century based upon similar considerations. It appeared *consistent* with facts, such as the existence and interconvertibility of known chemical compounds, but there was no *direct* structural evidence; the Horgans of the day might well have considered it to represent ironic science, if the term had occurred to them.[2] Nonetheless, these concepts proved extraordinarily productive in advancing the field of organic chemistry, though experimental methods for validating them did not come along for decades.

Indeed, *all* theoretical science starts from what is already known, and it would thus be fair to call all of it self-referential, within Ortega y Gasset's expanded understanding of "self." But that would remain true about any theory, until it leads to an experiment that tells us something new about the world, and is found to be either inconsistent with the result (in which case it would be just wrong, not ironic) or consistent. At that point the theory would no longer be only about itself—at least not the "self" under which it was formulated—because the world in which it is embedded—its circumstances—is no longer exactly the same. Ironic science ceases to be ironic when the world changes.

World Change and the End of Irony in Literature

As noted earlier, irony (in the amused/superior/detached/cynical/satirical sense of the term) pervades all three of the afore-mentioned novels—Jonathan Safran Foer's *Everything Is Illuminated*, José Saramago's *The Double*, and Arthur Phillips's *Prague*—and self-reference also plays a highly visible role. But the ironic tone disappears at significant points in the works. Here I will briefly summarize the novels, illustrate the authors' use (and abandonment) of irony, and suggest that they can thus be related to the ideas considered in the previous sections.

Everything Is Illuminated, the first novel by then 25-year-old Jonathan Safran Foer, tells the story of a young aspiring Jewish-American author (also named Jonathan Safran Foer), searching in 1997 for the remnants of his ancestral Ukrainian shtetl, Trachimbrod, and for a Ukrainian woman, Augustine, who helped his grandfather escape the Nazi invasion and liquidation. He engages a travel agency, which provides him with a translator, Alex; Alex's grandfather serves as driver, accompanied by a dog named Sammy Davis, Junior, Junior—in honor of his favorite singer. The work takes the form of a double narrative: a history of Trachimbrod, dating back to 1791, which is written by Jonathan but (mostly) told in an anonymous, omniscient voice; and Alex's account of the search. Self-referentiality is foregrounded by our knowledge that these narratives represent ongoing writing projects; the narrators exchange a series of post-search letters discussing their efforts (we only see Alex's letters to Jonathan, but those refer to the ones in the opposite direction). Each frequently urges the other to make changes—substantive as well as stylistic. Alex writes:

> We are being very nomadic with the truth, yes? The both of us? Do you think that this is acceptable when we are writing about things that occurred? If your answer is no, then why do you write about Trachimbrod and your grandfather in the manner that you do, and why do you command me to be untruthful? If your answer is yes, then this creates another question, which is if we are to be such nomads with the truth, why do we not make the story even more premium than life?
>
> (179)

Irony is at the forefront of both narrative lines, *almost* right up to the end. Jonathan's historic account of Trachimbrod reports a seemingly endless series of disasters—deaths, infant mortality, rapes, fatal accidents, pogroms—while maintaining an ironic, detached tone right from the opening account of the drowning—or possibly not—of Trachim, the shtetl's namesake:

> But sifting through the remains, they didn't find a body. For the next one hundred fifty years, the shtetl would host an annual contest to "find" Trachim, although a shtetl proclamation withdrew the reward in 1793—on Menasha's counsel that any ordinary corpse would begin to break apart after two years in water, so searching not only would be pointless but could result in rather offensive findings, or even worse, multiple rewards .... There were those who suspected that he was not pinned under the wagon but swept out to sea .... It's possible that he, or some part of him, washed up on the sands of the Black Sea, or in Rio, or that he made it all the way to Ellis Island.
>
> (14–15)

while the major ironic feature of Alex's narrative is his, shall we say, original English usage. Although he informs us that he "had performed recklessly well in my second year of English at university" (2), he manages at best an approximation thereof, with an over-reliance—at Jonathan's urging—on Roget:

> Dear Jonathan, I hanker for this letter to be good. Like you know, I am not first rate with English. In Russian my ideas are asserted abnormally well, but my second tongue is not so premium. I undertaked to input the things you counseled me to, and I fatigued the thesaurus you presented me, as you counseled me to, when my words appeared too petite, or not befitting. If you are not happy with what I have

performed, I command you to return it back to me. I will persevere
to toil on it until you are appeased.

(23)

Jonathan's ironical historic narrative comes to a dead halt at the Nazi
destruction of the town and its inhabitants. We learn about that instead
from Alex, translating a survivor's account, in a tone without any trace
of irony—straightforward, factual, chilling:

> "They made us in lines," she said. "They had lists .... They burned
> the synagogue." You cannot know how it felt to have to hear these
> things and then repeat them, because when I repeated them, I felt like
> I was making them new again .... "The General went down the line
> and told each man to spit on the Torah or they would kill his family."
> "This is not true," Grandfather said. "It is true," Augustine said, and
> she was not crying, which surprised me very much, but I understand
> now that she had found places for her melancholy that were behind
> more masks than only her eyes.
>
> (185)

For Jonathan the long, mostly awful history of Trachimbrod can be
viewed with amused detachment: it is just the "normal" world in which
its inhabitants lived. At one point Alex challenges that attitude: he calls
Jonathan a coward for "liv[ing] in a world that is 'once-removed'" (240).
But the Holocaust is something else altogether: it is world-changing, and
irony cannot be sustained. Alex undergoes a similar transformation, as
we see by comparing his accounts of encounters with a waitress before he
hears the survivor's story:

> "So just one mochaccino will be adequate," I told the waitress,
> who was a very beautiful girl with the most breasts I had ever seen
> .... "Would you like to do the Electric Slide with me at a famous
> discotheque tonight?" I asked the waitress. "Will you bring the
> American?" she asked. Oh, did this piss all over me! "He is a Jew,"
> I said .... "Oh," she said. "I have never seen a Jew before. Can I see
> his horns?" (It is possible that you will think she did not inquire this,
> Jonathan, but she did. Without a doubt, you do not have horns ....)
>
> (106–107)

and after:

> "You are returned," said the waitress when she witnessed us. "Back
> with the Jew," she said. "Shut your mouth," Grandfather said, and
> he did not say it in an earsplitting voice, but quietly, as if it were a
> fact that she should shut her mouth. "I am apologizing," she said. "It
> is not a thing," I told her, because I did not want her to feel inferior

for a small mistake, and also I could see her bosom when she bent forward. (For whom did I write that, Jonathan? I do not want to be disgusting anymore. And I do not want to be funny, either.)

(219)

Having grown up knowing next to nothing about the war and the fate of Ukrainian Jews, Alex finds that this "illumination" changes his world as well, and his narrative tone changes along with it.

José Saramago's *The Double* recounts the tale of Tertuliano Máximo Afonso, a high-school history teacher. Unlike *Everything Is Illuminated*, here the narrator is impersonal, but the element of self-reference is equally strong: he continually calls attention to himself and the fact that he *is* telling a tale:

> Sitting now on the bus that will drop him near the building where he has lived for the last six or so years, that is, ever since his divorce, Máximo Afonso, and we use the shortened version of his name here, having been, in our view, authorized to do so by its sole lord and master, but mainly because the word Tertuliano, having appeared so recently, only six lines previously, could do a grave disservice to the fluency of the narrative, anyway, as we were saying ....

(4)

In this passage, and throughout most of the book, the author reveals his attitude towards his central character: amused, superior, gently condescending—in sum, ironic, as Muecke notes: "the very act of being ironical implies an assumption of superiority" (31). Another example:

> Tertuliano Máximo Afonso is at home, he has a hesitant look on his face, not that this means very much, it isn't the first time it's happened, as he watches his will swing between spending time preparing something to eat, which generally means nothing more strenuous than opening a can and heating up the contents, or, alternatively, going out to eat in a nearby restaurant where he is known for his lack of interest in the menu, not because he is a proud, dissatisfied customer, he is merely indifferent, inattentive, reluctant to take the trouble to choose a dish among those set out in the brief and all-too-familiar list.

(7–8)

Things begin to change when he watches a mindless video and happens to spot a minor character who looks very much like him. At first he doesn't make too much of it:

> Apart from a few slight differences, he thought, especially the mustache, the different hairstyle, the thinner face, he's just like me. He felt

calmer now, the resemblance was, to say the least, astonishing, but that's all it was, and there's no shortage of resemblances in the world, twins for example, the really amazing thing would be that out of the six thousand million people on the planet there weren't two people exactly alike.

(17)

but then, realizing that the film was five years old:

Five years, he said again, and suddenly, the world gave another almighty shudder .... With trembling hands, he opened and closed drawers, pulled out envelopes full of negatives and photographs, scattered them over his desk, and, at last, found what he was looking for, a photo of himself, five years ago. He had a mustache, a different hairstyle, and his face was thinner.

(18)

Note how Saramago explicitly represents this epiphany as world-changing: "an almighty shudder." This is not atypical, as one reviewer commented:

In most of Saramago's novels, a major change occurs in the world: people go blind, or the Iberian peninsula detaches itself from the European mainland .... In dramatizing the aftermath of such changes, Saramago mercilessly satirizes those whose investment in the old status quo makes it impossible for them to adapt or even understand how obsolete their vision of the world has become.

(Parks 1999, 22)

However, Tertuliano Máximo Afonso is *not* one of those figures of authorial ridicule who refuse to adapt to a new world. He becomes obsessed with tracking down his double, whose name is António Claro; when they meet, they discover they are indeed perfect matches, down to the smallest detail (quality of voice, moles, a scar, born half an hour apart on the same day). From that point forward the narrative changes substantially, from a whimsical, leisurely, somewhat mocking examination of the foibles and missteps of a rather hapless antihero to a much terser tale of the tragic consequences of his quest. Claro tries to demonstrate his dominance by seducing Tertuliano Máximo Afonso's girlfriend, which ends in their being killed in an auto accident. But everyone thinks that the one who died is Tertuliano Máximo Afonso, who by then has changed dramatically from the indecisive person we first met. He opts to go on with life *as* Claro, with Claro's wife.

The culmination of his transformation—and the final end of irony—takes place at the very end, when Tertuliano Máximo Afonso answers a phone call (as Claro) and realizes how utterly his world has changed: *another* double is calling.

I've been looking for you for weeks, and I've finally found you .... Anyone seeing us together would swear that we were twins .... Tertuliano Máximo Afonso took a deep breath, then asked ... where can I meet you, It will have to be in some isolated spot, where there will be no witnesses .... He grabbed a bit of paper and scribbled, I'll be back, but did not sign it. Then he went into the bedroom and opened the drawer containing the pistol. He put the clip into the stock of the gun and transferred a cartridge into the chamber. He changed his clothes, clean shirt, tie, trousers, jacket, his best shoes. He stuck the pistol into his belt and left.

(324)

Prague is a story of a group of American expats looking for success in the newly liberated and newly capitalist Eastern Europe of the 1990s. As in the two previous books, a world-changing event is central to the plot; but here that event—the fall of the Iron Curtain and the end of the Soviet-dominated regime in Hungary and the other Warsaw Pact nations in 1989—has already taken place before the narrative commences. Furthermore, it did not really change these Americans' world all that much; it merely provided them with an opportunity to live much as they had always lived—hopefully even better—in a new place.

Accordingly, unlike the last two books, the ironic tone is basically uninterrupted throughout, starting with the title: the book is set entirely in Budapest. Prague is only occasionally mentioned as the place they *should* have gone: "Fifteen years from now people will talk about all the American artists and thinkers who lived in Prague in the 1990s. That's where real life is going on right now, not here" (14). (That opinion might well itself be ironic: it is offered as part of a game, invented by one of the main characters, called "Sincerity," in which players try to get their opponents to mistake authentic statements for lies, and *vice-versa*.) And it continues to the very end, when one of the characters, having given up on making a go of it in Budapest, takes a train to Prague (finally!):

He awakens, and there she is at last .... a single image exposed in a moment's glance: a land of spires and toy palaces and golden painted gates and bridges with sad-eyed statues peering out over misty black water ... and that fairy-tale castle floating above it, hovering unanchored by anything at all, a city where surely anything will be possible.

(367)

Only in a few passages does the narrative represent the points of view of actual Hungarians—for whom the world *did* change drastically—and *those* are often irony-free. One such describes the behavior of an elderly widower who is compelled to rent his previously state-subsidized Budapest apartment to one of the American protagonists and move to the countryside with his son; he is so distraught that he can't even take his family pictures with him:

[T]he father stood in front of his starkly empty armoire and simply stared at it .... after a moment the old man just went to the sofa and lay down on his stomach, his head tucked under his arm, turned away from the room .... Behind John, the old man was up and off the sofa, pulling something off the wardrobe's top shelf .... The old man held two framed pictures. After a deliberation, he placed one on top of the cable box and the other on the bedside table next to the lamp. He stretched his arms out to the two pictures, his fingers spread wide and his palms facing the frames, clearly to say: *Leave them like that.*

(23–25)

Such raw, honest portrayal of emotion is virtually absent from the vast majority of the novel.

In one scene Scott (one of the expats, who teaches English to a class of Hungarian academics and professionals) gives a lesson based on a newspaper column (written by Scott's brother) that claims their generation of Americans helped bring down the Berlin Wall and create a new society in Eastern Europe. Scott asks the class whether they think the column was meant to be taken seriously, which elicits an indictment of irony by Tibor, one of the students (who himself has a PhD in Hungarian literature):

His pupil proceeded: "It is my belief that irony is the tool of culture between high periods. It is the necessary fertilizer of the culture when it is, how does one—*mi az angolul, hogy parlagon hever?*" Zsófi, though entirely at a loss as to what Tibor was getting at, was the fastest with the *Magyar-Angol* dictionary. "To lie fallow," she reported proudly .... "Fallow. Yes," he began again. "American culture lies fallow now. There is nothing living, only things waiting. And the earth gives off only a smell. This smell, not pleasant, is irony. Like this newspaper writer. Very self-knowing." .... The rest of the class looked to Scott as the day's curriculum had unexpectedly delved deeply into agricultural questions .... "This is not good," insisted Zsófi. "It is a simple question, yes? Does he think it is true, he saved us from Russians by liking to watch MTV?"

(55)

This revealing passage exhibits *both* irony and sincerity, not to mention being self-referential about its own irony. The association of irony with a "fallow" period essentially captures my main point in this chapter: irony is how we see the world while we are, in effect, waiting for something to happen, for the world to change. Whether it is accompanied metaphorically by an unpleasant smell, as Scott's pupil proposes, is debatable. Rosenblatt and other 9/11 commentators seem to think so; perhaps Horgan would as well. But it can also be viewed as positive, helping us to live with a situation that is both static and (necessarily) imperfect: irony is something we really *can't* do without, as another contemporary novelist tells us:

> "It's the hardest addiction of all," said Patrick. "Forget heroin. Just try giving up irony, that deep-down need to mean two things at once, to be in two places at once, not to be there for the catastrophe of a fixed meaning."
>
> (St. Aubyn 2012, 60)

In much the same way, Horgan's concept of ironic, self-referential science is a myopic vision, a consequence of focusing on a frozen moment in time; it will be superseded when the world changes, along with the "self" that "self-referential" refers to. Such a development can always be anticipated with confidence, as the heading to both the first and last chapter of Jonathan's narrative in *Everything Is Illuminated* reminds us: "The Beginning of the World Often Comes" (8, 267). Science will not dissolve into irony, and we need have no concern about its end. To return to the Victor Hugo quote that closed Chapter 8, this time in (my own rather loose) translation: if science wants to find perpetual motion, it need only look to itself.

Notes

1 This chapter is based on a talk presented at the 2005 (Chicago) SLSA meeting.
2 A well-known example of such an attitude was German chemist Herman Kolbe's 1877 essay on structural theory, which included a harsh attack on Dutch chemist Jacobus van't Hoff:

> A Dr J. H. van't Hoff of the veterinary school at Utrecht finds, as it seems, no taste for exact chemical investigation. He has thought it more convenient to mount Pegasus (obviously loaned at the veterinary school) and to proclaim in his *La chimie dans l'éspace* how during his bold flight to the top of the chemical Parnassus, the atoms appeared to him to have grouped themselves throughout universal space.
>
> (Brock 1992, 262)

Kolbe went on to complain:

> It is one of the signs of the times that modern chemists hold themselves bound to consider themselves in a position to give an explanation for everything, and, when their knowledge fails them, to make use of supernatural (*sic*) explanations.
>
> (263)

Change "supernatural" to "ironic," and Horgan might have written that himself.

12 Conclusion

I gave a presentation on the subject of one of the preceding chapters at a meeting some years ago; during the ensuing discussion I was quite taken aback when a leading figure in the L&S community asked: *cui bono?* What are you trying to do, and whom do you think it might benefit? The form of the question was not unfamiliar to me: after I deliver a chemistry talk, perhaps about some mechanistic problem, often some industrial chemist in the audience demands (politely, to be sure) to be shown the practical value of the work. I typically reply by pointing out that fundamental knowledge, no matter how arcane it may appear, may always have future applicability. But I wasn't expecting such a challenge at SLSA! I stammered through something about how I was mainly doing it for myself, trying to see how ideas from nominally quite disparate areas that interested me might fit together; and that I believed such exploration could be useful to a broader community, but was not at that moment prepared to elucidate much on the latter point.

To a large extent, the present book is my belated attempt to provide a more coherent response. The case studies are intended to illustrate the connectionist approach that I introduced and (hopefully) justified in Chapter 4. Serendipity played a major role in their selection: most started with encountering a scientific paper that resonated in some way with one or more works of literature I had previously read, or the other way around, inspiring me to look more deeply into the nature and implications of that resonance. None of the studies is complete: to reiterate the point I made in Chapter 3, there is no such thing as "the whole story" about anything of interest. More intensive scholarship on these topics, as well as some of the specific works I have considered, would surely be rewarding. (Some of that has, of course, already been done by others.)

Furthermore, areas of overlap between the subjects of two (or more) chapters offer fruitful opportunities for further investigation. Several can already be seen in Chapters 5–7, which move broadly from code to simplicity to translation, showing how all of those three concepts work to establish linkages both between literature and science and with each other; there is still considerably more to be done along those lines. Other promising combinations include questions such as:

DOI: 10.4324/9781003197188-12

- to what degree and how does the impossibility of control in creating or modifying life (Chapter 10) follow from considerations of entropy and the second law (Chapter 8)?
- the notion of self-reference or reflexivity plays an important role in the discussions of both autopoietic thinking (Chapter 9; also see Hayles 1999, 131–159) and irony (Chapter 11); can those be related to one another in any productive way?
- questions of the relevance of autopoiesis to the natural origin of life were posed in Chapter 9; what about relevance to the artificial creations of life in Chapter 10? Or, for that matter, to "artificial life" itself, which Hayles discusses in terms of ideas from autopoiesis and systems theory (1999, 222–246)? Could concepts from autopoiesis help to distinguish, in any meaningful way, between artificial life (*e.g.*, a conscious computer) and what we might call artificially created *real* life?

Such questions amount to exploring connections between connections, much like (analogous to?) the comment about the desirability of "analogies between analogies" quoted in Chapter 4. This seems to me what Wilson's rather vaguely stated goal of "unifying knowledge" (also discussed in Chapter 4) is really about: approaching questions from as many directions as possible can give us more confidence in the answers. As the (fictional) Revd. Wicks Cherrycoke (what a name!) said:

[History must be] not a Chain of single Links, for one broken Link could lose us All,— rather, a great disorderly Tangle of Lines, long and short, weak and strong, vanishing into the Mnemonick Deep, with only their Destination in common.

(Pynchon 2004, 349)

This sort of entanglement is vital to *all* fields of human knowledge, not just history. A reviewer from the realm of music (a theme that has appeared several times in this book) expressed it thus:

What is the most important way to understand a piece of music, by a deep analysis of its content outside time, understanding the mechanics of its composition, or by a deep understanding of its cultural surroundings and the circumstances that brought it into being? .... Why cannot it be both/and rather than either/or? Each discipline, both deep analysis and rich cultural studies, adds to our understanding: great pieces ... are worth *every ounce of effort across different disciplines* that one might devote to them.

(Kenyon 2020, 55–56; my italics)

More generally, we need to avoid preconceptions of where answers are to be found. A historian of science, writing on experimental systems, commented:

Throughout the disciplines—the sciences, philosophy, sociology, anthropology, art history, and history of science— ... research is thus seen as an iterative process of groping—literally, *re*-search—that operates on the border between the known and the unknown. The basic problem lies in the fact that one does not precisely know what one *does not know*. What is at stake is the creation of new knowledge, and what is really new, is, by definition, unforeseeable.

(Rheinberger 2015, 169)

These last few quotes underline a theme that has arisen frequently in this book. In Chapter 3 I argued that nobody can tell a complete story about anything. I presented the relationships between metaphor/analogy and a connectionist model for brain function as a form of second-order, multi-level interconnectivity in Chapter 4; in Chapter 5 I highlighted another manifestation of such interconnectivity in *The Gold Bug Variations*, along with Powers's exhortation to embrace, not be frightened by, infinite possibility. I interpreted Stoppard as calling for never-ending questioning in *Arcadia* (Chapter 8); I devoted virtually an entire essay (Chapter 11) to the impossibility of reaching ends. For every problem we are interested in, there will always be aspects that extend beyond what may at first glance seem to be self-containing boundaries, in directions that we cannot anticipate *a priori*.

Morris Zapp, another fictional character (whose creator is a literary scholar as well as a novelist) whom we met briefly in Chapter 5, acknowledges that his goal of examining the novels of Jane Austen "from every conceivable angle .... So that when each commentary was written, there would be *nothing further to say*" was doomed to inevitable failure (Lodge 1984, 28–29, italics in original). Zapp takes the road (as did many of his real-life contemporaries) of abandoning that quest in favor of demonstrating its futility in terms of the inherent ambiguity of language. The alternative of continuing the pursuit—wherever it leads, always recognizing that the quarry will remain forever elusive—seems to me at least equally worthwhile. My connectionist approach to L&S, as exemplified in the case studies here, is just such an experimental program, in Rheinberger's sense.

As for the *cui bono* part of the question that opened this concluding chapter: in Chapter 1 I identified several audiences—general readers, literary scholars, and scientists—and suggested the potential benefits of attention to L&S for each group. Making the case for practicing scientists is perhaps the most challenging, so I will spend a little more time on that aspect, starting with my own experiences. In Chapter 6, I told of jumping to an erroneous conclusion in my early scientific career, and speculated that had I been more aware of the ideas discussed here, I *might* have been a little less hasty. While I can't be certain that my interest in L&S has indeed protected me from such mistakes, at least one subsequent experience seems to me suggestive of that. One of my postdocs showed me an

NMR spectrum (the same type of experiment that led to the earlier mistake) he had obtained from a reaction that should have resulted in only one or two products, and hence only one or two peaks; instead there were a couple dozen (Figure 12.1). He concluded there must have been *many* products, and we were misunderstanding the chemistry. This was after I had put together an earlier version of Chapter 6 for a presentation; and I recall thinking at the time that this could be another example of a misread coded message. Hence, it might be more productive to look for a technical issue that could cause an artifactual multiplicity of peaks, rather than trying to understand how the reaction could have led to multiple products. That turned out to be the correct interpretation; and that we reached it almost immediately owes at least *something*, I believe, to my recognizing a connection to the earlier episode via considerations of simplicity and coding.

Parallels between translation and representation of bonding and chemical structure have likewise proven useful to me, especially in the context of teaching. As I commented in Chapter 7, students can often get somewhat confused by going back and forth between the "language" of the different bonding models, and my realization that the problems they have are in many ways analogous to those presented to a translator has helped me come up with better explanations. I believe that other chemical educators could make use of these ideas as well. I gave a talk on the subject to an all-chemist audience at an American Chemical Society meeting, and it seemed to be received very positively (of course, I suppose they may have just wanted to be polite).

Engineering professor David Edwards offers a similar argument for cross-boundary collaborative thinking—what he calls "artscience"—in his book of that title (2008). His career underwent some changes, including a move from a chemical engineering department to a writing program, and he was struck by how they affected him:

> These shifts of perspective invigorated me. They also sensitized me to things—such as how my education had unwittingly narrowed the range of ideas I felt able to develop—I might not have noticed had I spent all my creative time in a single intellectual environment or culture. What separated these environments was a conceptual line .... I became enamored of these "unexplored" lines.

(2)

Edwards tells the stories of a number of "art/scientists" who have profited by taking that road, including a chemical engineer whose artistic background inspires him to formulate a new theory of fluid mixing (31–38), a musician who re-educates herself as an engineer and invents new methods of composition (22–31), and others. Not satisfied with merely advocating for his ideas, he established "laboratories" for the "translation of ideas" across interdisciplinary boundaries at Harvard and in Paris. Some

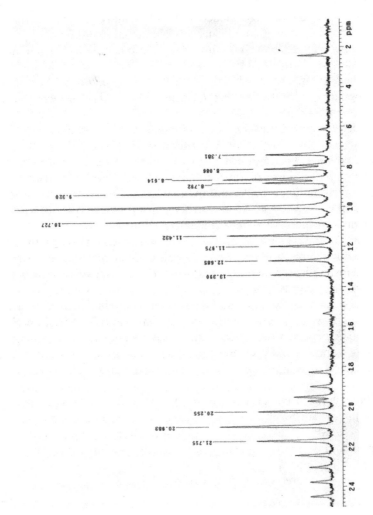

Figure 12.1 An NMR spectrum showing signals for phosphorus nuclei, which according to the expected chemistry should have exhibited a maximum of two peaks.

related L&S programs have been created within academic institutions (see Chapter 2), and of course SLSA itself represents such an initiative for *organized* collaborative interaction. Hopefully this will continue to be an expanding trend.

I shouldn't overstate claims for the utility of L&S; no doubt one could tell any number of counter-stories about those who have spread themselves too thinly by virtue of their broad and multiple interests, and accomplished less in their primary field than if they had been more single-minded. (Perhaps I'm one of them—who knows?) Indeed, many (probably most) of my highly successful chemist friends and colleagues *do* focus, laser-like, on their scientific pursuits, which can easily consume all of one's time and intellect.[1] I expect that's the main reason why SLSA has never attracted more than a handful of active scientists. Edwards's Paris venture seems to have primarily generated art exhibits, rather than scientific advances; and in any case, it lasted only a few years, from 2007 to 2014, before closing (Paris Art n.d.).

Despite such risks, I believe that taking an interest in L&S—even one well short of full-fledged active participation—can be beneficial in many ways: for breaking out of routine thinking habits; for lowering interdisciplinary barriers both inside and outside of academia; and above all, for fostering the recognition that *everyone* pursuing knowledge and understanding is fundamentally engaged in the same activity. As the late physician-essayist Lewis Thomas said, pleading for mutual respect:

> I intend to take a stand in the middle of what seems to me a muddle, hoping to confuse the argument by showing that there isn't really any argument in the first place. To do this, I must try to show that there is in fact a solid middle ground to stand on, a shared common earth beneath the feet of all the humanists and all the scientists, a single underlying view of the world that drives all scholars, whatever their discipline—whether history or structuralist criticism or linguistics or quantum chromodynamics or astrophysics or molecular genetics. There is, I think, such a shared view of the world. It is called *bewilderment*.
>
> (Thomas 1983, 157)

That, I suggest, is the essence of L&S: to keep our minds open to unfamiliar modes of thought, and to new ways of addressing our problems, wherever they may be found.

Note

1 At the beginning of this book I mentioned my early exposure to the claim that generalists have all the advantage over specialists; a full-book-length argument endorsing that position cites some supporting data: "Scientists and members of the general public are about equally likely to have artistic hobbies, but

scientists inducted into the highest national academies are much more likely to have avocations outside of their vocation" (Epstein 2019, 32–33). Perhaps that's so—data is data!—but my (admittedly limited) personal experience tends to point in the opposite direction: more correlation between greater scientific recognition and greater single-minded pursuit thereof.

Appendix 1: Some Details of the Chemistry

The concept of stereochemistry—that the way the atomic components of a chemical substance are arranged in space is an important aspect of its nature—has its origins in the early 19th century, when it was recognized that certain crystalline solids have the ability to rotate polarized light. The latter is produced by passing ordinary light through a plate of a polarizing material that, by virtue of its structure, restricts the direction of the light wave vibrations to lie within a particular plane. (Many sunglasses use this principle.) The existence and magnitude of rotation are readily measurable by passing the transmitted light through a second plate and adjusting the angle of its plane of polarization relative to that of the first until light is transmitted, using an instrument called a polarimeter or polariscope.

Subsequently it was discovered that not only crystals but also *solutions* of some compounds can effect such rotation, a property termed *optical activity*. In the 1870s this behavior was interpreted in terms of the detailed geometry of organic (carbon-based) molecules, which frequently consist of a central carbon atom with four attached groups (called substituents). Those substituents are arranged so as to point to the corners of a tetrahedron centered around the carbon atom. If they are all different, then there are *two* possible structures for the same composition, as we can see in Figure A.1. The mirror image of one is *not* identical to—cannot be superimposed upon—itself, just as one's left and right hands are not superimposable. Such molecules are said to be *chiral* (from the Greek word for hand), and the two mirror-image forms are called optical isomers or enantiomers. (There are other stereochemical situations that can lead to chirality, but we do not need to consider those here.) Solutions of the two pure enantiomers will rotate the plane of polarization in opposite directions. In contrast, a solution containing equal amounts of the two enantiomers, called a *racemic* mixture, will not rotate the plane of polarization at all. For a brief (and much more entertaining) account of this history—which includes connections to Jewish tradition!—see Hoffmann and Schmidt (1997).

As discussed in Chapter 6, in an S_N2 reaction the incoming group approaches from the direction *opposite* to the leaving group, such that the

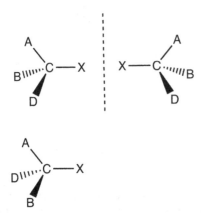

Figure A.1 Top: reflection of the structure of a compound containing a carbon
center and four different substituents in a mirror (dotted line).
Bottom: the right-hand structure from the top line, rotated so as to
show two of the substituents (A and X) in the same orientation as the
left-hand structure. Recalling that solid and dashed wedges represent
bonds pointing in front and in back of the page, respectively, we can
see that the two mirror images are different: the positions of the other
two substituents B and D do *not* coincide. (Note that if there were
only three different substituents, *e.g.*, if B and D were identical, the
mirror images *would* be superimposable, and the compound would
not be chiral.)

geometry about the carbon is effectively turned inside out. We noted that
this cannot be tested for a methyl compound: it can be seen by inspection
of Figure A.1 that if even two of the substituents are identical, there will
not be any difference between the isomers, so inversion has no detectable
consequences. But use of a more complex alkyl halide does permit such
a study, as seen in Figure 6.1. Classically such experiments have been
carried out using polarimetry, and indeed, one such proposed a concerted
mechanism for oxidative addition on the basis of a purported observation of retention of stereochemistry (Pearson and Muir 1970). However,
since the direction of rotation of light for two *different* compounds
with opposite geometric orientations need *not* be opposite, that conclusion relied upon an (unjustified) auxiliary assumption. We looked
for an unequivocal way to test inversion that would not depend on any
assumptions about auxiliary experiments, and found one in nuclear magnetic resonance, or NMR.

It is worth spending a little time on NMR, which is not only the main
tool in the vast body of organic chemistry research but also the basis
of a ubiquitous medical diagnostic procedure. (The following discussion
is *extremely* simplified; any reference I might give will probably be too
technical for much of the readership of this book, but anyone interested

in going a little more deeply into the history and principles might look at Goldenberg (2016) and references cited therein.) Like any form of spectroscopy, it depends upon a system that can exist in two (or more) states of differing energy. If the energy difference between two states corresponds to the frequency of some form of electromagnetic radiation (the relation between energy E and frequency ν is given by the Planck–Einstein relation $\Delta E = h\nu$, where h is Planck's constant), then the system can absorb energy from the radiation, resulting in some observable phenomenon. This is a state of resonance, completely analogous to that described in the very first chapter, where an unstruck bell takes some energy from the sound waves produced by a struck one and sounds itself. The most familiar form involves so-called electronic excitation, the energy of which often corresponds to frequencies in the range of visible light; that energy absorption is responsible for most of the colors we see.

In the case of NMR, it has been known since early in the 20th century that most atomic nuclei behave as though they are spinning (to what extent this represents a literal or metaphorical description continues to be debated), resulting in their exhibiting a magnetic moment. If placed in a magnetic field, then (like a compass needle in the earth's magnetic field), they too will exist in one of two (or more) energy states, aligned with or against the field (Figure A.2). Those energy differences are *very* small, corresponding to radio wave frequencies. In the 1940s it was demonstrated that excitation of a proton in a hydrogen atom to

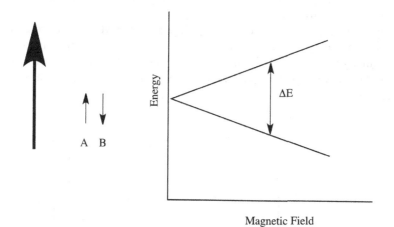

Magnetic Field

Figure A.2 The basis of NMR spectroscopy. On the left, the magnetic moment of a nucleus may be oriented either parallel (small arrow A) or opposed (small arrow B) to an applied magnetic field (heavy arrow). The graph on the right shows the relationship between the difference in energy of the two orientations, ΔE, and the strength of the magnetic field. If a radio wave is applied, energy will be absorbed when the frequency matches (is in resonance with) that difference.

the higher-energy state could be detected by placing a suitable sample in a magnetic field, generating a radio signal at a chosen frequency, and then varying the strength of the field, thus changing the energy difference between the two states. At the point where the magnetic field value is such that the Planck–Einstein relation is satisfied—when the frequency corresponding to the energy difference of the two states of the proton is the same as (in *resonance* with) that of the applied radio signal—the sample absorbs electromagnetic energy, resulting in a detectable signal.

At first this appeared to be little more than an interesting curiosity, but within a few years two key properties were discovered and exploited. First, the energy of resonance is *not* the same for all protons, but varies considerably (and fairly predictably) with the chemical environment of the hydrogen atom in question. Second, the magnetic moment of each nucleus generates its own magnetic field—tiny compared with the externally applied field, but sufficient to perturb (slightly) the field "felt" by its neighbors (an effect known as spin-spin coupling). As a consequence, the NMR spectrum of an organic compound can serve practically as a fingerprint for identification purposes. Figure A.3 shows one example, that of ethanol, where the signals for the two different hydrocarbon groups, CH_3 and CH_2, resonate at different positions (this is called a chemical shift difference); and the two signals are split into multiplets of three and four lines, respectively (generally n protons close enough to interact will split a signal into n + 1 lines). It seems to me entirely appropriate to think of such spectra as coded messages: the spectrometer tells us about the molecular structure, but does so in the form of an array of lines, which needs, in effect, to be decoded.

Before going on to discuss the application of NMR to my research problem, I briefly digress to consider the medical application—magnetic resonance imaging or MRI—which will be more familiar, and hence perhaps of greater interest, to readers. It might be first noted that the absence

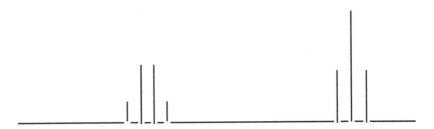

Figure A.3 Idealized NMR spectrum of ethanol, CH_3CH_2OH. The signal for the CH_3 group is split into three lines (a triplet) by interactions with the two protons of the CH_2 group; analogously, the signal for the latter is split into four lines (a quartet). (The signal for the OH group is not shown.)

of an "N" in the latter is not for technical reasons—the basic scientific principles are the same as in NMR—but it was thought desirable to avoid any undesirable connotations of including "nuclear" in the name. MRI uses a somewhat different method of exciting the nuclei (as does contemporary NMR instrumentation): a *pulse* of radio wave energy is applied, sufficiently strong to excite *all* the nuclei, thus changing the magnetization of the entire sample, and the behavior of the resulting signal is followed over time. That time-dependent evolution can be processed mathematically (known as a Fourier Transform) to separate out each of the different signals in the sample.

MRI usually focuses on one particular feature of that evolution, known as *relaxation*: the rate at which an excited nucleus returns to the more favored equilibrium, or lower-energy, state. Most of the hydrogen atoms in the body are in the form of water, and it turns out that the relaxation rate of protons in water is highly sensitive to the environment in which it is found—blood, protein, bone, tumors, *etc.* How is spatial imaging achieved? Unlike standard NMR, for which the magnetic field needs to be perfectly uniform throughout the sample to obtain sharp signals, MRI makes use of an *inhomogeneous* magnetic field, whose local strength is in effect moved relative to the part of the body being examined; the signals and how they change with orientation are recorded. The resulting (*huge*) data set undergoes computer processing to produce an image revealing the presence and location of these different regions. As an essentially non-invasive procedure that is just as sensitive to soft tissue as bone (unlike X-rays), this has become an extremely valuable diagnostic tool.

Now, back to NMR and my research problem. It was also discovered in the 1950s that the *magnitude* of the interaction causing spin-spin splitting is sensitive to geometry, and hence, NMR can be used to study stereochemistry—the very experiment we wanted to carry out—with a suitably designed molecular probe. Our initial choice was the compound shown in Figure A.4, where we planned to look at the NMR spectrum of the fluorine nucleus (which behaves much the same as a proton with respect to chemical shifts and spin-spin splitting). Such molecules, consisting of six-membered rings of carbon atoms with substituents, nearly always prefer the non-planar geometry shown (called a "chair" structure). In such structures there are two possible orientations for each substituent, either up/down (as for three of the four H atoms shown in the starting compound), called *axial*, or angularly sideways (as for the Br and F atoms), called *equatorial*. Virtually always, the larger substituents preferably occupy equatorial positions. In the starting structure both Br and F can be thus accommodated; likewise for M and F in the product resulting from retention. But if inversion takes place, that is no longer possible: the structure will readjust to place M (which is much larger, with its associated additional groups) in an equatorial site, forcing F to be axial.

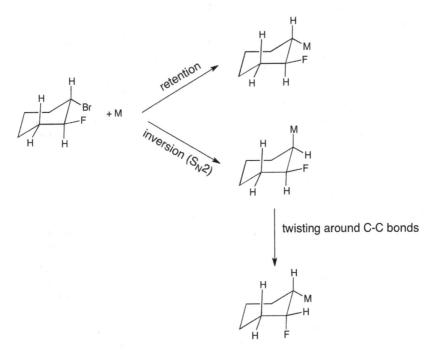

Figure A.4 Structure of the molecule selected to probe the stereochemical outcome of oxidative addition of a carbon-bromine bond to a metal
center, where M represents the metal along with additional attached
groups. The unlabeled vertices of the hexagon all represent CH_2
groups. The F atom ends up in an equatorial position with retention,
but an axial position with inversion (after the structure twists to put
the M group in its preferred equatorial position).

It was also well known that there should be strong (and roughly equal)
coupling of the F nucleus to i) the H nucleus attached to the same C;
and ii) any H on an adjacent C *if* the C-F and C-H bonds are oriented
oppositely (at 180°) to one another. In contrast, F and H nuclei on adjacent carbons with small angles between the C-F and C-H bonds would
give much smaller coupling. By examining Figure A.4, it can be seen to
be possible to predict—with confidence—that a reaction with inversion
would result in a F signal showing strong coupling to three H nuclei—*i.e.*,
a quartet—while retention would give one with only one such strong
coupling—a doublet. There would also be weaker coupling to (many)
additional protons, so each of the "lines" would actually be broadened
envelopes of lines. Those expected spectra, and what we actually observed,
were shown in Figure 6.2.

Appendix 2: Suggestions for Further Reading

The following list—like the book as a whole—makes no attempt at any sort of comprehensive coverage; in particular, it does not include works focused on any specific theme or topic. Rather, it offers some leading general references that may be helpful to readers wishing to delve more broadly and/or deeply into L&S. (Some of these are also included in the Works Cited section, but are repeated here for convenience.)

1 Bibliographic compilations

Schatzberg, Walter, Ronald A. Waite, and Jonathan K. Johnson. 1987. *The Relations of Literature and Science: An Annotated Bibliography of Scholarship, 1880–1980*. New York: The Modern Language Association of America.

Scholnick, Robert J. 1992. "Bibliography: American Literature and Science through 1989." In Scholnick 1992, 251–272.

Annual bibliographies for 1989–2000 appeared in the SLSA journal *Configurations* (locations may be found by searching at https://muse.jhu.edu/journal/36); and for the years 2001–2004 and 2011, on the SLSA website, www.litsciarts.org/bibliography/

An L&S "timeline" that includes selected important works may be found in Meyer 2018, *xiv–xxii*.

2 Encyclopedias, handbooks, texts

Wilson, David L., and Zack Bowen. 2001. *Science and Literature: Bridging the Two Cultures*. Gainesville, FL: University Press of Florida.

Gossin, Pamela, Ed. 2002. *Encyclopedia of Literature and Science*. Westport, CT: Greenwood Press.

Cartwright, John H., and Brian Baker. 2005. *Literature and Science: Social Impact and Interaction*. Santa Barbara, CA: ABC-CLIO.

Clarke, Bruce, and Manuela Rossini, Eds. 2011. *The Routledge Companion to Literature and Science*. London: Routledge.

Meyer, Steven, Ed. 2018. *The Cambridge Companion to Literature and Science*. Cambridge: Cambridge University Press.

3 Essay collections: monographs

Garvin, Harry R, ed. 1983. *Science and Literature*. Lewisburg, PA: Bucknell University Press.

Levine, George. 1987, Ed. *One Culture: Essays in Science and Literature*. Madison, WI: University of Wisconsin Press.

Amrine, Frederick, Ed. 1989. *Literature and Science as Modes of Expression*. Dordrecht: Kluwer Academic Publishers.

Slade, Joseph W., and Judith Yaross Lee, Eds. 1990. *Beyond the Two Cultures*. Ames, IA: Iowa State University Press.

Peterfreund, Stuart, Ed. 1990. *Literature and Science: Theory and Practice*. Boston, MA: Northeastern University Press.

Scholnick, Robert J., Ed. 1992. *American Literature and Science*. Lexington, KY: The University Press of Kentucky.

4 Essay collections: focused journal issues

Annals of Scholarship 1986, Volume 4 (1): "Science and the Imagination."

University of Hartford Studies in Literature 1987, Volume 19 (1).

Journal of Literature and Science 2017, Volume 10 (1): "The State of the Unions Part 1." (online only: www.literatureandscience.org/volume-10-issue-1-2017/)

Configurations 2018, Volume 26 (3): "The State of the Unions Part 2."

Works Cited

Adams, Henry. 1910. *A Letter to American Teachers of History*. Washington, DC: Press of J. H. Furst Company.

Alexander, Daryl R. 1996. "Word for Word: A Gathering of Skeptics." *New York Times*, July 7, Section 4: 7.

Amato, Ivan. 2007. "Researchers Unearth Another Stratum of Meaning in the Genetic Code." *Chemical & Engineering News* 85 (January 22): 38–40.

Anzalone, Andrew V., Peyton B. Randolph, Jessie R. Davis, Alexander A. Sousa, Luke W. Koblan, Jonathan M. Levy, Peter J. Chen, Christopher Wilson, Gregory A. Newby, Aditya Raguram, and David R. Liu. 2019. "Search-and-Replace Genome Editing Without Double-Strand Breaks or Donor DNA." *Nature* 576 (October 21): 149–157.

Arendt, Hannah. 1968. "Introduction: Walter Benjamin: 1892–1940." In *Walter Benjamin, Illuminations*, translated by Harry Zohn, edited by Hannah Arendt. New York: Schocken Books. Kindle.

Armstrong, Paul B. 2013. *How Literature Plays with the Brain: The Neuroscience of Reading and Art*. Baltimore, MD: Johns Hopkins University Press.

Arnold, Matthew. 1882. "Literature and Science." *The Nineteenth Century* 12 (August): 216–230.

Asimov, Isaac. (1956) 1969. "The Last Question." In *Nine Tomorrows*, 170–183. New York: Fawcett Crest.

Atwood, Margaret. 2003. *Oryx and Crake*. London: Virago Press.

———. 2009. *The Year of the Flood*. New York: Doubleday.

———. 2011. *In Other Worlds: SF and the Human Imagination*. New York: Anchor Books.

———. 2013. *MaddAddam*. New York: Random House.

Avery, Oswald T., Colin M. Macleod, and Maclyn McCarty. 1944. "Studies on the Chemical Nature of the Substance Inducing Transformation of Pneumococcal Types: Induction of Transformation by a Desoxyribonucleic Acid Fraction Isolated from *Pneumococcus* Type III." *Journal of Experimental Medicine* 79 (2): 137–158.

Bard, Allen J. 1996. "The Antiscience Cancer." *Chemical & Engineering News* 74 (April 22): 5.

Baudrillard, Jean. 1994. *Simulacra and Simulation*, translated by Sheila Faria Glaser. Ann Arbor, MI: University of Michigan Press.

Baumgartner, Emily. 2018. "As D.I.Y. Gene Editing Gains Popularity, 'Someone Is Going to Get Hurt.'" *New York Times*, May 15: D1.

Beadle, George and Muriel Beadle. 1966. *The Language of Life: An Introduction to the Science of Genetics.* Garden City, NY: Doubleday.

Beer, Gillian. 1983. *Darwin's Plots: Evolutionary Narrative in Darwin, George Eliot and Nineteenth-Century Fiction.* Cambridge, UK: Cambridge University Press.

———. 1990. "Forging the Missing Link: Interdisciplinary Stories." In *Companion to the History of Modern Science*, edited by Robert C. Olby, Geoffrey N. Cantor, John R. R. Christie and M. Jonathon S. Hodge, 783–798. London: Routledge.

———. 1996. "Science and Literature." In *Open Fields: Science in Cultural Encounter*, 115–145. Oxford, UK: Clarendon Press.

Begley, Sharon. 1997. "The Science Wars." *Newsweek*, April 21: 54–57.

Bénard, Marc. 1979. "Molecular Orbital Analysis of the Metal-Metal Interaction in Some Carbonyl-Bridged Binuclear Complexes." *Inorganic Chemistry* 18 (10): 2782–2785.

Benfey, O. Theodore. 1952. "Prout's Hypothesis." *Journal of Chemical Education* 29 (2): 79–81.

Berg, Paul, David Baltimore, Sydney Brenner, Richard O. Roblin, III, and Maxine F. Singer. 1975. "Summary Statement of the Asilomar Conference on Recombinant DNA Molecules." *Proceedings of the National Academy of Science* 72 (6): 1981–1984.

Berman, Sabina. 1998. *Bubbeh*, translated by Andrea G. Labinger. New York: Latin American Literary Review Press.

Bernal, John Desmond. 1967. *The Origin of Life.* New York: World Publishing Company.

Besser, Steven. 2017. "How Patterns Meet: Tracing the Isomorphic Imagination in Contemporary Neuroculture." *Configurations: A Journal of Literature, Science, and Technology* 25 (4): 415–445.

Black, Max. 1962. *Models and Metaphors.* Ithaca, NY: Cornell University Press.

———. 1993. "More About Metaphor." In Ortony 1993a, 19–41.

Blackmond, Donna G., Christopher R. McMillan, Shailesh Ramdeehul, Andrea Schorm, and John M. Brown. 2001. "Origins of Asymmetric Amplification in Autocatalytic Alkylzinc Addition." *Journal of the American Chemical Society* 123 (41): 10103–10104.

Bloor, David. (1976) 1991. *Knowledge and Social Imagery.* 2nd Ed. Chicago: University of Chicago Press.

Bono, James J. 1990. "Science, Discourse, and Literature: The Role/Rule of Metaphor in Science." In Peterfreund 1990, 59–89.

———. 2010. "Making Knowledge: History, Literature, and the Poetics of Science." *Isis* 101 (3): 555–559.

Borek, Ernest. 1965. *The Code of Life.* New York: Columbia University Press.

Borges, Jorge Luis. 1964a. "The Analytical Language of John Wilkins." In *Other Inquisitions (1937–1957)*, translated by Ruth L. C. Simms, 101–105. Austin, TX: University of Texas Press.

———. 1964b. "Death and the Compass," translated by Donald A. Yates. In *Labyrinths: Selected Stories and Other Writings*, edited by Donald A. Yates and James E. Irby, 76–87. New York: New Directions.

———. 1964c. "The Library of Babel," translated by James E. Irby. In *Labyrinths: Selected Stories and Other Writings*, edited by Donald A. Yates and James E. Irby, 51–58. New York: New Directions.

Boyd, Richard. 1993. "Metaphor and Theory Change: What is 'Metaphor' a Metaphor for?" In Ortony 1993a, 481–532.

Bradley, John S., Dan E Connor, Jay A. Labinger, David Dolphin, and John A. Osborn. 1972. "Oxidative Addition to Iridium(I): A Free-Radical Process." *Journal of the American Chemical Society* 94(11): 4043–4044.

Brainard, Jeffrey. 2019. "CRISPR's First Clinical Success?" *Science* 366 (November 22): 930.

Brock, William H. 1992. *The Norton History of Chemistry*. New York: W. W. Norton.

Bronowski, Jacob. 1965. *Science and Human Values*. New York: Harper Torchbooks.

Brown, Theodore L. 2003. *Making Truth: Metaphor in Science*. Urbana, IL: University of Illinois Press.

Bruni, John. 2011. "Thermodynamics." In Clarke and Rossini 2011, 226–237.

Bunge, Mario. 1996. "In Praise of Intolerance to Charlatanism in Academia." In Gross *et al.* 1996, 96–115.

Bursten, Bruce E., and Roger H. Cayton, 1986. "Electronic Structure of Piano-Stool Dimers. 3. Relationships Between the Bonding and Reactivity of the Organically Bridged Iron Dimers $[CpFe(CO)]_2(\mu\text{-}CO)(\mu\text{-}L)$ (L = CO, CH_2, $C=CH_2$, CH^+)." *Journal of the American Chemical Society* 108 (26), 8241–8249.

Cardiff University. n.d. "Cardiff ScienceHumanities." https://cardiffsciencehumanities.org/#content-wrapper. Accessed January 24, 2020.

Carroll, Joseph. 1995. *Evolution and Literary Theory*. Columbia, MO: University of Missouri Press.

———. 2005. "Human Nature and Literary Meaning: A Theoretical Model Illustrated with a Critique of Pride and Prejudice." In Gottschall and Wilson 2005b, 76–106.

Carroll, Joseph, Dan P. McAdams, and Edward O. Wilson, Eds. 2016. *Darwin's Bridge: Uniting the Humanities and Sciences*. Oxford, UK: Oxford University Press.

Cartwright, John H., and Brian Baker. 2005. *Literature and Science: Social Impact and Interaction*. Santa Barbara, CA: ABC-CLIO.

Cech, Thomas R. 1999. "Science at Liberal Arts Colleges: A Better Education?" *Daedalus* 128 (1): 195–216.

Chabon, Michael. 2000. *The Amazing Adventures of Kavalier and Clay*. New York: Picador.

Charlwood, Catherine. 2018. "[Don't] Leave the Science Out: An Argument for the Necessary Pairing of Cognition and Culture." *Configurations: A Journal of Literature, Science, and Technology* 26 (3): 303–310.

Chock, Pwen Boon, and Jack Halpern. 1966. "Kinetics of the Addition of Hydrogen, Oxygen, and Methyl Iodide to Some Square-Planar Iridium (I) Complexes." *Journal of the American Chemical Society* 88 (15): 3511–3514.

Ciardi, John. 1954. *Dante Alighieri, The Inferno: A Verse Rendering for the Modern Reader*. New York: New American Library.

Clark, Jeff. 2019. "An Escape into Fiction." *Science* 366 (November 15): 918.

Clarke, Bruce. 2011. "Systems Theory." In Clarke and Rossini 2011, 214–225.

Clarke, Bruce, and Manuela Rossini, Eds. 2011. *The Routledge Companion to Literature and Science*. London: Routledge.

Clayton, Jay. 2013. "The Ridicule of Time: Science Fiction, Bioethics, and the Posthuman." *American Literary History* 25 (2): 317–343.

Cohen, I. Bernard. 1993. "Analogy, Homology and Metaphor in the Interactions Between the Natural Sciences and the Social Sciences, Especially Economics." In *Non-Natural Social Science: Reflecting on the Enterprise of More Heat than Light*, edited by Neil de Marchi, 7–44. Durham, NC: Duke University Press.

Colebrook, Claire. 2004. *Irony*. London: Routledge.

Collini, Stefan. 1998. "Introduction." In Snow (1959) 1998, *vi–lxxi*.

Collins, Harry M. 1981a. "Son of Seven Sexes: The Social Destruction of a Physical Phenomenon." *Social Studies of Science* 11: 33–62

———. 1981b. "Stages in the Empirical Programme of Relativism." *Social Studies of Science* 11: 3–10.

———. 1992. *Changing Order: Replication and Induction in Scientific Practice*. 2nd Ed. Chicago: University of Chicago Press.

———. 1994. "A Strong Confirmation of the Experimenters' Regress." *Studies in the History and Philosophy of Science* 25 (3): 493–503.

Collins, Harry, Robert Evans, and Martin Weinel. 2017. "STS as Science or Politics?" *Social Studies of Science* 47 (4): 580–586.

Collins, Harry, and Trevor Pinch. 1993. *The Golem: What Everyone Should Know About Science*. Cambridge, UK: Cambridge University Press.

Collins, Harry, and Steven Yearley. 1992. "Journey into Space." In *Science as Practice and Culture*, edited by Andrew Pickering, 369–389. Chicago, IL: The University of Chicago Press.

Crane, Mary Thomas. 2015. "Analogy, Metaphor and the New Science: Cognitive Science and Early Modern Epistemology." In Zunshine 2015c, 103–114.

Crews, Frederick. 2005. "Forward from the Literary Side." In Gottschall and Wilson 2005b, *xiii–xv*.

Cyranoski, David, and Heidi Ledford. 2018. "International Outcry Over Genome-Edited Baby Claim." *Nature* 563 (November 29): 607–608.

Dante Alighieri. 1317. *The Inferno*. www.mediasoft.it/dante/pages/danteinf.htm. Accessed April 16, 2014.

Davies, Jamie A. 2018. *Synthetic Biology: A Very Short Introduction*. Oxford, UK: Oxford University Press.

Dawkins, Richard. 1998. *Unweaving the Rainbow: Science, Delusion and the Appetite for Wonder*. Boston: Houghton Mifflin Co.

de Montaigne, Michel. 1834. *Essais de Michel de Montaigne, Avec Des Notes de Tous les Commentateurs*. Paris: Chez Lefevre. Viewed on Google Books, September 18, 2019.

Denbigh, Kenneth G., and Jonathan S. Denbigh. 1985. *Entropy in Relation to Incomplete Knowledge*. Cambridge, UK: Cambridge University Press.

de Vrieze, Jop. 2017. " 'Science Wars' Veteran Has a New Mission." *Science* 358 (October 13): 159.

Dihal, Kanta. 2017. "On Science Fiction as a Separate Field." *Journal of Literature and Science* 10 (1): 32–36.

Dillon, Sarah. 2018. "On the Influence of Literature on Science." *Configurations: A Journal of Literature, Science and Technology* 26 (3): 311–316.

Divan, Aysha, and Janice A. Royds. 2016. *Molecular Biology: A Very Short Introduction*. Oxford, UK: Oxford University Press.

Dobzhansky, Theodosius. 1973. "Nothing in Biology Makes Sense Except in the Light of Evolution." *American Biology Teacher* 35 (3): 125–129.

Doudna, Jennifer A., and Samuel H. Sternberg. 2017. *A Crack in Creation: Gene Editing and the Unthinkable Power to Control Evolution*. Boston: Houghton Mifflin Harcourt.

Drake, Stillman. 1960. *The Assayer*. Translation of *Il Saggiatore* by Galileo Galilei. www.stanford.edu/~jsabol/certainty/readings/Galileo-Assayer.pdf. Accessed April 16, 2014.

Dreistadt, Roy. 1968. "An Analysis of the Use of Analogies and Metaphors in Science." *The Journal of Psychology* 68: 97–116.

Droge, Abigail. 2017. "Teaching Literature and Science in Silicon Valley." *Journal of Literature and Science* 10 (1): 58–64.

Dyson, Freeman. 1998. "Is God in the Lab?" *New York Review of Books* 45 (May 28). www.nybooks.com/articles/1998/05/28/is-god-in-the-lab/

———. 1999. *Origins of Life*. Cambridge, UK: Cambridge University Press.

———. 2011. "How to Dispel Your Illusions." *New York Review of Books* 58 (December 22): 40–43.

Eco, Umberto. (1980) 1983. *The Name of the Rose*, translated by William Weaver. New York: Houghton Mifflin.

Eddington, Arthur S. 1948. *The Nature of the Physical World*. New York: Macmillan.

Edelman, Gerald. 1987. *Neural Darwinism*. New York: Basic Books.

———. 2006. *Second Nature: Brain Science and Human Knowledge*. New Haven, CT: Yale University Press.

Edelman, Gerald, and Giulio Tononi. 2000. *A Universe of Consciousness: How Matter Becomes Imagination*. New York: Basic Books.

Edwards, David. 2008. *Artscience: Creativity in the Post-Google Generation*. Cambridge, MA: Harvard University Press.

Eilenberger, Wolfram. 2020. *Time of the Magicians: Wittgenstein, Benjamin, Cassirer, Heidegger, and the Decade That Reinvented Philosophy*, translated by Shaun Whiteside. New York: Penguin Press.

ELINAS. n.d. "Center for Literature and Natural Sciences." http://elinas.fau.de/index.en.html. Accessed April 9, 2019.

Elliott, Robin. 2006. "Margaret Atwood and Music." *University of Toronto Quarterly* 75 (Summer): 821–832.

Epstein, David. 2019. *Range: Why Generalists Triumph in a Specialized World*. New York: Riverhead Books.

Falkoff, Rebecca. 2021. "Lipstick Vitalism: On the Beauty of a Different Modernity in Primo Levi's *Periodic Table*." *Configurations: A Journal of Literature, Science, and Technology* 29 (1): 53–72.

Faraday, Michael. (1857) 2008. "Letter to James Clerk Maxwell, 13 November 1857." In *The Correspondence of Michael Faraday, Vol. 5*, edited by Frank A. L. James, 304–306. London: The Institution of Engineering and Technology.

Feyerabend, Paul. 1975. *Against Method*. London: New Left Books.

Flint, Anthony. 1994. "Science Isn't Immune to Cultural Critique." *Boston Globe*, November 15: 1, 28.

Foer, Jonathan Safran. 2002. *Everything Is Illuminated*. New York: HarperCollins.

Forman, Paul. 1971. "Weimar Culture, Causality, and Quantum Theory, 1918–1927: Adaptation by German Physicists and Mathematicians to a Hostile Intellectual Environment." *Historical Studies in the Physical Sciences* 3: 1–115.

Foucault, Michel. 1970. *The Order of Things: An Archaeology of the Human Sciences*. New York: Random House.

Franklin, Allan. 1994. "How to Avoid the Experimenters' Regress." *Studies in the History and Philosophy of Science* 25 (3): 463–491.

Franklin, Allan, and Harry Collins. 2016. "Two Kinds of Case Study and a New Agreement." In *The Philosophy of Historical Case Studies, Boston Studies in the Philosophy of Science*, edited by T. Sauer and R. Scholl, 95–121. Dordrecht: Springer.

Frayn, Michael. 1998. "Postscript." In *Copenhagen*, 95–132. New York: Anchor Books.

Freudenberger, Nell. 2019. *Lost and Wanted*. New York: Alfred A. Knopf.

Friedrich, Otto. 1989. *Glenn Gould: A Life with Variations*. New York: Random House.

Fuller, Steve. 1994. "Can Science Studies Be Spoken in a Civil Tongue?" *Social Studies of Science* 24 (1): 143–168.

Galilei, Galileo. 1623. *Il Saggiatore*. http://scholar.google.com/scholar?q=galileo+il+saggiatore&btnG=&hl=en&as_sdt=0%2C5. Accessed April 16, 2014.

Galison, Peter, and David J. Stump, Eds. 1996. *The Disunity of Science: Boundaries, Contexts, and Power*. Stanford, CA: Stanford University Press.

Garner, Dwight. 2019. "In Michel Houellebecq's 'Serotonin,' the Provocative Beat Goes On (and On)." *New York Times*, November 18: C4.

Gass, Willam H. 1970. "In Terms of the Toenail: Fiction and the Figures of Life." In *Fiction and the Figures of Life*, 55–76. New York: Alfred A. Knopf.

Geary, James. 2012. *I Is an Other: The Secret Life of Metaphor and How It Shapes the Way We See the World*. New York: Harper Perennial.

Gentner, Dedre, and Michael Jeziorski. 1993. "The Shift from Metaphor to Analogy in Western Science." In Ortony 1993a, 447–480.

Geraghty, James. 1996. "Adenomatous Polyposis Coli and Translational Medicine." *The Lancet* 348 (August 17): 422.

Gibbs, Raymond W., Jr., Ed. 2008. *The Cambridge Handbook of Metaphor and Thought*. Cambridge, UK: Cambridge University Press.

Gleick, James. 1987. *Chaos: Making a New Science*. New York: Viking.

Goldenberg, David P. 2016. *Principles of NMR Spectroscopy: An Illustrated Guide*. Mill Valley, CA: University Science Books.

Goldstein, Rebecca Newberger. 2005. "Gödel and the Nature of Mathematical Truth." *Edge*, June 8. www.edge.org/3rd_culture/goldstein05/goldstein05_index.html. Accessed October 24, 2019.

Golinski, Jan. 1998. *Making Natural Knowledge: Constructivism and the History of Science*. Cambridge, UK: Cambridge University Press.

Goodstein, David L. 1989. "Richard P. Feynman, Teacher." *Physics Today* 42 (2): 70–75.

Gopnik, Adam. 2000. *Paris to the Moon*. New York: Random House.

Gordin, Michael D. 2015. *Scientific Babel: How Science Was Done Before and After Global English*. Chicago: University of Chicago Press.

Gossin, Pamela, Ed. 2002. *Encyclopedia of Literature and Science*. Westport, CT: Greenwood Press.

Gottschall, Jonathan, and David Sloan Wilson. 2005a. "Introduction: Literature—a Last Frontier in Human Evolutionary Studies." In Gottschall and Wilson 2005b, *xvii–xxvi*.

———, Eds. 2005b. *The Literary Animal: Evolution and the Nature of Narrative*. Evanston, IL: Northwestern University Press.

Gould, Stephen J. 1999. *Rocks of Ages: Science and Religion in the Fullness of Life*. New York: Ballantine Publishing Group.

———. 2003. *The Hedgehog, the Fox, and the Magister's Pox: Mending the Gap Between Science and the Humanities*. Cambridge, MA: Harvard University Press.

Graham, Loren R. 1964. "A Soviet Marxist View of Structural Chemistry: The Theory of Resonance Controversy." *Isis* 55 (1): 20–31.

Green, Jennifer C., Malcolm L. H. Green, and Gerard Parkin. 2012. "The Occurrence and Representation of Three-Centre Two-Electron Bonds in Covalent Inorganic Compounds." *Chemical Communications* 48 (94): 11481–11503.

Greene, Brian. 1999. *The Elegant Universe: Superstrings, Hidden Dimensions, and the Quest for the Ultimate Theory*. New York: Norton.

Griffiths, Devin. 2016. *The Age of Analogy: Science and Literature Between the Darwins*. Baltimore, MD: Johns Hopkins University Press.

———. 2018. "Darwin and Literature." In Meyer 2018a, 62–80.

Grosholz, Emily R., and Roald Hoffmann. 2012. "How Symbolic and Iconic Language Bridge the Two Worlds of the Chemist." In *Roald Hoffmann on the Philosophy, Art, and Science of Chemistry*, edited by Jeffrey Kovac and Michael Weisberg, 222–241. Oxford, UK: Oxford University Press.

Gross, Barry R. 1996. "Flights of Fancy: Science, Reason, and Common Sense." In Gross *et al.* 1996, 79–86.

Gross, Paul R., and Norman Levitt. 1994. *Higher Superstition: The Academic Left and Its Quarrels with Science*. Baltimore, MD: The Johns Hopkins University Press.

Gross, Paul R., Norman Levitt, and Martin W. Lewis, Eds. 1996. *The Flight from Science and Reason*. New York: New York Academy of Sciences.

Guston, David H., Ed Finn, and Jason Scott Robert, Eds. 2017. *Frankenstein: Annotated for Scientists, Engineers, and Creators of All Kinds*. Cambridge, MA: The MIT Press.

Haack, Susan. 2003. *Defending Science—Within Reason: Between Scientism and Cynicism*. New York: Prometheus Books.

Hacking, Ian. 1992. "Statistical Language, Statistical Truth and Statistical Reason: The Self-Identification of a State of Scientific Reasoning." In *The Social Dimensions of Science*, edited by Ernan McMullin, 130–157. Notre Dame, IN: University of Notre Dame Press.

Hagendijk, Rob. 1990. "Structuration Theory, Constructivism, and Scientific Change." In *Theories of Science in Society*, edited by Susan E. Cozzens and Thomas F. Gieryn, 43–66. Bloomington, IN: Indiana University Press.

Hager, Thomas. 1995. *Force of Nature: The Life of Linus Pauling*. New York: Simon & Schuster.

Hall, Louisa. 2019. "Uncertainty Principles." *New York Times Book Review*, April 7: 1: 20.

Hamilton, Sheryl N. 2003. "Traces of the Future: Biotechnology, Science Fiction, and the Media." *Science Fiction Studies* 30: 267–282.

Hamner, Everett. 2017. *Editing the Soul: Science and Fiction in the Genome Age*. University Park, PA: The Pennsylvania State University Press.

Hawbaker, Neil A., and Donna G. Blackmond. 2019. "Energy Threshold for Chiral Symmetry Breaking in Molecular Self-Replication." *Nature Chemistry* 11 (10): 957–962.

Hawks, John. 2016. "Neanderthal Humanities." In Carroll *et al.* 2016, 89–100.

Hayles, N. Katherine. 1990a. *Chaos Bound: Orderly Disorder in Contemporary Literature and Science*. Ithaca, NY: Cornell University Press.

———. 1990b. "Self-Reflexive Metaphors in Maxwell's Demon and Shannon's Choice." In Peterfreund 1990, 209–237.

———. 1991. "Introduction: Complex Dynamics in Literature and Science." In *Chaos and Order*, edited by N. Katherine Hayles, 1–33. Chicago: Chicago University Press.

———. 1994. "Deciphering the Rules of Unruly Disciplines: A Modest Proposal for Literature and Science." In *Literature and Science*, edited by Donald Bruce and Anthony Purdy, 25–48. Amsterdam: Editions Rodopi.

———. 1999. *How We Became Posthuman: Virtual Bodies in Cybernetics, Literature, and Informatics*. Chicago: University of Chicago Press.

———. 2001. "Desiring Agency: Limiting Metaphors and Enabling Constraints in Dawkins and Deleuze/Guattari." *SubStance* 30 (1/2): 144–159.

Heinlein, Robert. (1941) 1962. "By His Bootstraps." In *The Menace from Earth*, 49–115. New York: New American Library.

Heisenberg, Werner. 1958. "The Representation of Nature in Contemporary Physics." *Daedalus* 87 (3): 95–108.

Hemingway, Ernest. 1940. *For Whom the Bell Tolls*. New York: Scribner.

Herman, Luc and Lernout, Geert. 1998. "Genetic Coding and Aesthetic Clues: Richard Powers's *Gold Bug Variations*." *Mozaic: A Journal for the Interdisciplinary Study of Literature* 31 (4): 151–164.

Hesse, Mary B. 1966. *Models and Analogies in Science*. Notre Dame, IN: University of Notre Dame Press.

———. 1980. "The Explanatory Function of Metaphor." In *Revolutions and Reconstructions in the Philosophy of Science*, 111–124. Bloomington, IN: Indiana University Press.

History of Science Society. n.d. "History of the Society." https://hssonline.org/about/history-of-the-society/. Accessed April 15, 2019.

Hoffmann, Roald. 2012. "Unstable." In *Roald Hoffmann on the Philosophy, Art, and Science of Chemistry*, edited by Jeffrey Kovac and Michael Weisberg, 39–44. Oxford, UK: Oxford University Press.

Hoffmann, Roald, Vladimir I. Minkin, and Barry K. Carpenter. 1996. "Ockham's Razor and Chemistry." *Bulletin de la Société Chimique de France* 133 (2): 117–130.

Hoffmann, Roald, and Shira Leibowitz Schmidt. 1997. "You Must Not Deviate to the Right or the Left." In *Old Wine New Flasks: Reflections on Science and Jewish Tradition*, 79–121. New York: W. H. Freeman.

Hofstadter, Douglas R. 1980. *Gödel, Escher, Bach: An Eternal Golden Braid*. New York: Vintage Books.

———. 1995. *Fluid Concepts and Creative Analogies: Computer Models of the Fundamental Mechanisms of Thought*. New York: Basic Books.

———. 1997. *Le Ton Beau de Marot: In Praise of the Music of Language*. New York: Basic Books.

———. 2009a. *That Mad Ache*. Translation of *La Chamade*, by Françoise Sagan. New York: Basic Books.

———. 2009b. *Translator, Trader: An Essay on the Pleasantly Pervasive Paradoxes of Translation*. New York: Basic Books.

Hofstadter, Douglas R., and Gary McGraw. 1995. "Letter Spirit: Esthetic Perception and Creative Play in the Rich Microcosm of the Roman Alphabet." In Hofstadter 1995, 407–466.

Hofstadter, Douglas R., and Emmanuel Sander. 2013. *Surfaces and Essences: Analogy as the Fuel and Fire of Thinking*. New York: Basic Books.

Hogan, Patrick Colm. 2015. "On Being Moved: Cognition and Emotion in Literature and Film." In Zunshine 2015c, 237–256.

Holton, Gerald. 2004. "Robert K. Merton: 4 July 1910–23 February 2003." *Proceedings of the American Philosophical Society* 148 (4): 505–517.

Horgan, John. 1996. *The End of Science: Facing the Limits of Knowledge in the Twilight of the Scientific Age.* Reading, MA: Helix Books.

———. 2004. "The End of Science Revisited." *Computer* (January): 37–43.

———. 2015. *The End of Science: Facing the Limits of Knowledge in the Twilight of the Scientific Age.* 2nd Ed. New York: Basic Books.

Houellebecq, Michel. 2000. *The Elementary Particles,* translated by Frank Wynne. New York: Vintage International.

Huxley, Aldous. 1963. *Literature and Science.* New York: Harper & Row.

Huxley, Thomas H. 1882. "Science and Culture: An Address Delivered at the Opening of Sir Josiah Mason's Science College, at Birmingham, on the 1st of October 1880." In *Science and Culture and Other Essays,* 1–23. London: Macmillan.

Ihde, Aaron J. 1964. *The Development of Modern Chemistry.* New York: Harper and Row.

Ingram, V. M. 1957. "Gene Mutations in Human Haemoglobin: The Chemical Difference Between Normal and Sickle Cell Haemoglobin." *Nature* 180 (August 17): 326–328.

Irwin, Aisling. 1994 "Sociology Row Erupts at BA." *Times Higher Education Supplement* (September 16): 4.

James, Anthony A. 2005. "Gene Drive Systems in Mosquitoes: Rules of the Road." *Trends in Parasitology* 21 (2): 64–67.

Japp, Francis R. 1898. "Science and Vitalism." *Nature* 58 (September 8): 452–460.

Jardine, Nick, and Marina Frasca-Spada. 1997. "Splendors and Miseries of the Science Wars." *Studies in History and Philosophy of Science* 28 (2): 219–235.

Jemmis, Eluvathingal D., Allan R. Pinhas, and Roald Hoffmann. 1980. "$Cp_2M_2(CO)_4$—Quadruply Bridging, Doubly Bridging, Semibridging, or Nonbridging?" *Journal of the American Chemical Society* 102 (8), 2576–2585.

Johnson, Phillip E. 1998. "Afterword: How to Sink a Battleship." In *Mere Creation: Science, Faith & Intelligent Design,* edited by William A. Dembski, 446–453. Downers Grove, IL: InterVarsity Press.

Jones, Louis B. 1991. "Bach Would've Liked This Molecule." Review of *The Gold Bug Variations,* by Richard Powers. *New York Times Book Review* (August 25): 9–10.

Kahneman, Daniel. 2011. *Thinking, Fast and Slow.* New York: Farrar, Straus and Giroux.

Kakutani, Michiko. 2000. "Books of the Times: Unsparing Case Studies of Humanity's Vileness." Review of *The Elementary Particles,* by Michel Houellebecq. *New York Times* (November 10): E47.

———. 2001. "The Age of Irony Isn't Over After All: Assertions of Cynicism's Demise Belie History." *New York Times* (October 9): E1, E5.

Kandel, Eric R. 2018. *The Disordered Mind: What Unusual Brains Tell Us About Ourselves.* New York: Farrar, Straus and Giroux.

Kay, Lily E. 2000. *Who Wrote the Book of Life? A History of the Genetic Code.* Stanford, CA: Stanford University Press.

Keller, Evelyn Fox. 1995. "Science and Its Critics." *Academe* (September–October): 10–15.

Kelly, Katherine E., and William W. Demastes. 1994. "The Playwright and the Professors: An Interview with Tom Stoppard." *South Central Review* 11 (4): 1–14.

Kenyon, Nicholas. 2020. "What Matters in Music?" Review of *Cursed Questions on Music and Its Social* Practice, by Richard Taraskin. *New York Review of Books* 67 (December 17): 55–57.

Koertge, Noretta, Ed. 1998. *A House Built on Sand: Exposing Postmodernist Myths About Science.* New York: Oxford University Press.

Koestler, Arthur. n.d. "Humour." *Encyclopedia Britannica*, online edition. www.britannica.com/topic/humor. Accessed April 5, 2019.

Kuhn, Thomas S. (1962) 1970. *The Structure of Scientific Revolutions.* 2nd Ed. Chicago: University of Chicago Press.

———. 1993. "Metaphor in Science." In Ortony 1993a, 531–542.

Labinger, Jay A. 1993. Review of *The Golem: What Everyone Should Know About Science*, by Harry Collins and Trevor Pinch. *Engineering and Science* 57 (1), 39–40.

———. 1995a. "Encoding an Infinite Message: Richard Powers's *The Gold Bug Variations.*" *Configurations: A Journal of Literature, Science, and Technology* 3 (1): 79–93.

———. 1995b. "Science as Culture: A View from the Petri Dish." *Social Studies of Science* 25 (2): 285–306.

———. 1996a. "Metaphoric Usage of the Second Law: Entropy as Time's (Double-Headed) Arrow in Tom Stoppard's *Arcadia.*" *The Chemical Intelligencer* 2 (4): 30–36.

———. 1996b. Review of *The End of Science*, by John Horgan. *Engineering and Science* 60 (4), 28–29.

———. 1997. "The Science Wars and the Future of the American Academic Profession." *Daedalus* 126 (4): 201–220.

———. 2002. "Bond-Stretch Isomerism: A Case Study of a Quiet Controversy." *Comptes Rendus Chimie* 5 (4): 235–44.

———. 2014. "Does Cyclopentadienyl Iron Dicarbonyl Dimer Have a Metal-Metal Bond? Who's Asking?" *Inorganica Chimica Acta* 424: 14–19.

———. 2015. "Tutorial on Oxidative Addition." *Organometallics* 34 (20), 4784–4795.

———. 2017. "Where Are the Scientists in Literature and Science?" *Journal of Literature and Science* 10 (1): 65–69.

Labinger, Jay A., Robyn J. Braus, David Dolphin, and John A. Osborn. 1970. "Oxidative Addition of Alkyl Halides to Iridium(I) Complexes." *Journal of the Chemical Society D: Chemical Communications* (10): 612–613.

Labinger, Jay A., and Harry Collins, Eds. 2001. *The One Culture? A Conversation About Science.* Chicago: University of Chicago Press.

Labinger, Jay A., and Stephen J. Weininger. 1995. "Conversing with Outsiders." Review of *Higher Superstition*, by Paul R. Gross and Norman Levitt. *Chemical & Engineering News* 73 (January 9): 27–28.

LabLit.com. n.d. "The Culture of Science in Fiction and Art." www.lablit.com. Accessed April 9, 2019.

Lacan, Jacques. 1978a. *Le Séminaire*, Book I, edited by Jacques-Alain Miller. Paris: Éditions du Seuil.

————. 1978b. *Le Séminaire*, Book II, edited by Jacques-Alain Miller. Paris: Éditions du Seuil.

Lakoff, George, and Mark Johnson. 1980. *Metaphors We Live By*. Chicago: University of Chicago Press.

Landy, Joshua. 2015. "Mental Calisthenics and Self-Reflexive Fiction." In Zunshine 2015c, 559–580.

Laplace, Pierre-Simon. (1814) 1998. *Philosophical Essay on Probabilities*, translated (from the 1825 edition) by Andrew I. Dale. New York: Springer-Verlag.

Latour, Bruno. 2004. "Why Has Critique Run out of Steam? From Matters of Fact to Matters of Concern." *Critical Inquiry* 30 (Winter): 225–248.

Latour, Bruno, and Steve Woolgar. 1986. *Laboratory Life: The Construction of Scientific Facts*. 2nd Ed. Princeton, NJ: Princeton University Press.

Lawler, Andrew. 1996. "Support for Science Stays Strong." *Science* 272 (May 31): 1256.

Layzer, David. 1975. "The Arrow of Time." *Scientific American* 233 (December): 56–69.

Leavis, Frank R. 1962. "Two Cultures? The Significance of C. P. Snow." *The Spectator* (March 9): 90–101.

Levi, Primo. [1984] 2015. "Asymmetry and Life," translated by Alessandra Bastagli and Francesco Bastagli. In *The Complete Works of Primo Levi*, edited by Ann Goldstein, Vol. 3, 2657–2667. New York: Liveright Publishing Corp.

————. [1985] 2015. "Preface to *Other People's Trades*," translated by Antony Shugaar. In *The Complete Works of Primo Levi*, edited by Ann Goldstein, Vol. 3, 2013–2014. New York: Liveright Publishing Corp.

Lévy-Leblond, Jean-Marc. 1993. "The Mirror, the Beaker and the Touchstone, or, What Can Literature Do for Science?" *SubStance* 71/72: 7–26.

Lewontin, Richard C. 2001. "In the Beginning Was the Word." Review of *Who Wrote the Book of Life*, by Lily Kay. *Science* 291 (February 16), 1263–1264.

————. 2014. "The New Synthetic Biology: Who Gains?" *New York Review of Books* 61 (May 8). www.nybooks.com/articles/2014/05/08/new-synthetic-biology-who-gains/.

Livingston, Ira. 2006. *Between Science and Literature: An Introduction to Autopoetics*. Urbana, IL: University of Illinois Press.

Lodge, David. 1984. *Small World*. New York: Macmillan.

————. 2004. "Goodbye to All That." Review of *After Theory*, by Terry Eagleton. *New York Review of Books* 51 (9), 39–43.

Longfellow, Henry Wadsworth. 1865. *Dante Alighieri, The Inferno*, translated by H. W. Longfellow. www.fullbooks.com/Dante-s-Inferno1.html. Accessed April 16, 2014.

Luisi, Pier Luigi. 2016. *The Emergence of Life: From Chemical Origins to Synthetic Biology*. Cambridge, UK: Cambridge University Press.

Lynch, Michael. 2001. "Is a Science Peace Process Necessary?" In Labinger and Collins 2001, 48–60.

Lyons, Thomas R., and Allan D. Franklin. 1973. "Thomas Pynchon's 'Classic' Presentation of the Second Law of Thermodynamics." *Bulletin of the Rocky Mountain Modern Language Association* 27 (4): 195–204.

Manchester, Keith. 2007. "Stereochemistry and Vitalism." *South African Journal of Science* 103: 68–70.

Markley, Robert. 2018. "As If: The Alternative Histories of Literature and Science." *Configurations: A Journal of Literature, Science, and Technology* 26 (3): 259–268.

Marshall, Michael. 2020. *The Genesis Quest: The Geniuses and Eccentrics on a Journey to Uncover the Origin of Life on Earth*. Chicago: University of Chicago Press.

Martínez, Guillermo. 2012. *Borges and Mathematics*, translated by Andrea G. Labinger. West Lafayette, IN: Purdue University Press.

Marx, William. 2018. *The Hatred of Literature*, translated by Nicholas Elliott. Cambridge, MA: Harvard University Press.

Massing, Michael. 2019. "Are the Humanities History?" *New York Review Daily* (April 2). www.nybooks.com/daily/2019/04/02/are-the-humanities-history/

Maturana, Humberto R. 1980. "Introduction." In Maturana and Varela 1980, *xi–xxx*.

Maturana, Humberto R., and Francisco J. Varela. 1980. "Autopoiesis: The Organization of the Living." In *Autopoiesis and Cognition: The Realization of the Living*, 73–140. Dordrecht: D. Reidel.

McKinney, William J. 1998. "When Experiments Fail: Is 'Cold Fusion' Science as Normal?" In Koertge 1998, 133–150.

Meierhenrich, Uwer. 2008. *Amino Acids and the Asymmetry of Life Caught in the Act of Formation*. Berlin: Springer.

Meyer, Steven, Ed. 2018a. *The Cambridge Companion to Literature and Science*. Cambridge, UK: Cambridge University Press.

———. 2018b. "Futures Past and Present: Literature and Science in an Age of Whitehead." In Meyer 2018a, 255–274.

———. 2018c. "Introduction." In Meyer 2018a, 3–21.

Milburn, Colin. 2010. "Modifiable Futures: Science Fiction at the Bench." *Isis* 101 (3): 560–569.

Miller, Arthur M. 2000. "Metaphor and Scientific Creativity." In *Metaphor and Analogy in the Sciences*, edited by Fernand Hallyn, 147–164. Dordrecht: Kluwer Academic Publishers.

Mingers, John. 1991. "The Cognitive Theories of Maturana and Varela." *Systems Practice* 4 (4): 319–337.

MLA. 1937. Proceedings of the Modern Language Association of America 52: 1349–1373.

——— 1938. Proceedings of the Modern Language Association of America 53: 1340–1363.

——— 1939. Proceedings of the Modern Language Association of America 54: 1356–1381.

——— 1943. Proceedings of the Modern Language Association of America 58 (4): 5–33.

——— 1980. Proceedings of the Modern Language Association of America 95 (6): 939–1056.

——— 1985. Proceedings of the Modern Language Association of America 100 (6): 867–1019.

——— 2015. "Literature and Science Studies Forum." http://english.artsci.wustl.edu/litsci. Accessed April 9, 2019 but no longer available.

Moats, Michael. 2012. "The *Infinite Jest Liveblog*: What Happened, Pt. 2." www.fictionadvocate.com/2012/09/19/the-infinite-jest-liveblog-what-happened-pt-2/. Accessed February 16, 2021.

Montgomery, Scott L. 2000. *Science in Translation: Movements of Knowledge Through Cultures and Time*. Chicago: University of Chicago Press.

Monti, Enrico. 2006. "Dwelling upon Metaphors: The Translation of William Gass's Novellas." *Nordic Journal of English Studies* 5 (1): 118–132.

Mooney, Chris. 2005. *The Republican War on Science*. New York: Basic Books.

Morgan, Mary S., and M. Norton Wise. 2017. "Narrative Science and Narrative Knowing. Introduction to Special Issue on Narrative Science." *Studies in History and Philosophy of Science Part A* 62: 1–5.

Muecke, Douglas Colin. 1969. *The Compass of Irony*. London: Methuen.

Mulisch, Harry. 2001. *The Procedure*, translated by Paul Vincent. New York: Penguin Books.

Müller, Ingo. 2007. *A History of Thermodynamics: The Doctrine of Energy and Entropy*. Berlin: Springer Verlag.

Nabokov, Vladimir. 1964. "Commentary." In *Eugene Onegin: A Novel in Verse. Translated from the Russian, with a Commentary, by Vladimir Nabokov*, Vol. 1. New York: Pantheon.

Nash, Richard. 2017. "A Sense of Belonging: The Place of Literature and Science in a More Ecologically Alert Academy." *Journal of Literature and Science* 10 (1): 70–74.

Newman, Daniel Aureliano. 2018. "Narrative: Common Ground for Literature and Science?" *Configurations: A Journal of Literature, Science and Technology* 26 (3): 277–282.

Nicolson, Marjorie Hope. 1965. "Resource Letter SL-1 on Science and Literature." *American Journal of Physics* 33 (3): 175–183.

Nirenberg, Marshall. 1977. "The Genetic Code." In *Nobel Lectures in Molecular Biology, 1933–1975*, edited by David Baltimore, 335–358. New York: Elsevier North-Holland.

Ochoa, Severo. 1962. "Enzymatic Mechanisms in the Transmission of Genetic Information." In *Horizons in Biochemistry*, edited by Michael Kasha and Bernard Pullman, 153–166. New York: Academic Press.

Olson, Randy. 2015. *Houston, We Have a Narrative: Why Science Needs Story*. Chicago: University of Chicago Press.

Oreck, Alden. n.d. "Modern Jewish History: The Golem." *Jewish Virtual Library*. www.jewishvirtuallibrary.org/the-golem. Accessed November 12, 2019.

Oreskes, Naomi, and Eric M. Conway. 2010. *Merchants of Doubt: How a Handful of Scientists Obscured the Truth on Issues from Tobacco Smoke to Global Warming*. London: Bloomsbury Press.

Oreskes, Naomi, Kristin Shrader-Frechette, and Kenneth Belitz. 1994. "Verification, Validation, and Confirmation of Numerical Models in the Earth Science." *Science* 263 (February 4): 641–646.

Ortega y Gasset, José. (1914) 2004, "Meditaciones del Quijote." In *Obras Completas*, Vol. I. Madrid: Tauruse/Fundación Ortega y Gasset.

Ortony, Andrew, Ed. 1993a. *Metaphor and Thought*. 2nd Ed. Cambridge, UK: Cambridge University Press.

———. 1993b. "Metaphor, Language, and Thought." In Ortony 1993a, 1–16.

Otis, Laura. 2015. "The Value of Qualitative Research for Cognitive Literary Studies." In Zunshine 2015c, 505–524.

Otto, Shawn Lawrence. 2016. *The War on Science: Who's Waging It, Why It Matters, What We Can Do About It*. Minneapolis, MN: Milkweed Editions.

Ozick, Cynthia. 1997. *The Puttermesser Papers*. New York: Vintage International.

Padian, Kevin. 2018. "Narrative and 'Anti-Narrative' in Science: How Scientists Tell Stories, and Don't." *Integrative and Comparative Biology* 58 (6): 1224–1234.

Pais, Abraham. 1982. *Subtle Is the Lord: The Science and the Life of Albert Einstein*. Oxford, UK: Oxford University Press.

Paris Art. n.d. "*Le Laboratoire (Lieu Fermé).* www.paris-art.com/lieux/le-laboratoire/. Accessed February 3, 2020.

Park, Robert L. 2000. "Voodoo Science and the Belief Gene." *Skeptical Inquirer* (September/October): 24–29.

Parks, Tim. 1999. "Sightgeist." Review of *Blindness*, by José Saramago. *New York Review of Books* 46 (February 18): 22–25.

Pauling, Linus. 1939. *The Nature of the Chemical Bond and the Structure of Molecules and Crystals*. Ithaca, NY: Cornell University Press.

Pauling, Linus, Harvey A. Itano, S. J. Singer, and Ibert C. Wells. 1949. "Sickle Cell Anemia, a Molecular Disease." *Science* 110 (November 25): 543–548.

Paulson, William. 2001. *Literary Culture in a World Transformed: A Future for the Humanities*. Ithaca, NY: Cornell University Press.

Paz Soldán, Edmundo. 2003. *El Delirio de Turing*. Buenos Aires: Alfaguara.

———. 2006. *Turing's Delirium*, translated by Lisa Carter. Boston: Houghton Mifflin.

Pearson, Ralph G., and Warren R. Muir. 1970. "Mechanism of Oxidative Addition Reactions. Retention of Configuration in the Reaction of Alkyl Halides." *Journal of the American Chemical Society* 92 (18): 5519–5520.

Pennisi, Elizabeth. 2007. "DNA Study Forces Rethink of What It Means to Be a Gene." *Science* 316 (15 June): 1556–1557.

Perkowitz, Sidney. 2002. "Quantum Physics." In Gossin 2002, 364–366.

Peterfreund, Stuart, Ed. 1990. *Literature and Science: Theory and Practice*. Boston: Northeastern University Press.

Phillips, Arthur. 2002. *Prague*. New York: Random House.

Philosophy of Science Association. n.d. "History of the Association." https://philsci.org/history_of_the_association.php. Accessed April 15, 2019.

Pigliucci, Massimo. 2016. "The Limits of Consilience and the Problem of Scientism." In Carroll *et al.* 2016, 247–264.

Pinker, Steven. 2002. *The Blank Slate: The Modern Denial of Human Nature*. New York: Viking.

———. 2018. "The Intellectual War on Science." *Chronicle of Higher Education* (February 13). www.chronicle.com/article/The-Intellectual-War-on/242538.

Platt, J. R. 1962. "A 'Book Model' of Genetic Information-Transfer in Cells and Tissues." In *Horizons in Biochemistry*, edited by Michael Kasha and Bernard Pullman, 167–187. New York: Academic Press.

Plotnitsky, Arcady. 2005. "Science and Narrative." In *Routledge Encyclopedia of Narrative Theory*, edited by David Herman, Manfred Jahn, and Marie-Laure Ryan, 514–518. Abingdon, UK: Routledge.

Polizzotti, Mark. 2018. *Sympathy for the Traitor: A Translation Manifesto*. Cambridge, MA: The MIT Press.

Porush, David. 1991. "Fictions as Dissipative Structures: Prigogine's Theory and Postmodernism's Roadshow." In *Chaos and Order*, edited by N. Katherine Hayles, 54–84. Chicago: Chicago University Press.

Powers, Richard. 1985. *Three Farmers on Their Way to a Dance*. New York: William Morrow.

——. 1988. *Prisoner's Dilemma*. New York: William Morrow.

——. 1991. *The Gold Bug Variations*. New York: William Morrow.

——. 2014. *Orfeo*. New York: W. W. Norton.

Powner, Matthew W., and John D. Sutherland. 2011. "Prebiotic Chemistry: A New *Modus Operandi*." *Philosophical Transactions of the Royal Society B*, 366: 2870–2877.

Prigogine, Ilya, and Isabelle Stengers. 1984. *Order Out of Chaos: Man's New Dialogue with Nature*. New York: Bantam Books.

Pynchon, Thomas. (1960) 1984. "Entropy." In *Slow Learner*, 79–98. Boston: Little, Brown.

——. 1966. *The Crying of Lot 49*. London: Picador.

——. 1984. "Introduction." In *Slow Learner*, 1–24. Boston: Little, Brown.

——. 2004. *Mason and Dixon*. New York: Picador.

Quinn, Anthony. 2000. "One Thinks, the Other Doesn't." Review of *The Elementary Particles*, by Michel Houellebecq. *New York Times* (November 19), Section 7: 8.

Remmel, Ariana. 2020. "Searching the Galaxy for Signs of Life." *Chemical & Engineering News* 98 (November 30): 31–35.

Rheinberger, Hans-Jörg. 2015. "Difference Machines: Time in Experimental Systems." *Configurations: A Journal of Literature, Science, and Technology* 23 (2): 165–176.

Richardson, Alan. 2018. "Literary Studies and Cognitive Science." In Meyer 2018a, 207–222.

Rifkin, Jeremy. 1980. *Entropy: A New World View*. New York: Viking Press.

Roberts, David. 1999. "Self-Reference in Literature." In *Problems of Form*, edited by Dirk Backer, 27–45. Stanford, CA: Stanford University Press.

Robinson, Charles E. 2017. "Introduction." In Guston *et al.* 2017, *xxii–xxxv*.

Roll-Hansen, Nils. 2000. "The Application of Complementarity to Biology: From Niels Bohr to Max Delbrück." *Historical Studies in the Physical and Biological Sciences*. 30 (2): 417–442.

Rosenbaum, Thane. 2002. *The Golems of Gotham*. New York: HarperCollins.

Rosenblatt, Roger. 2001. "The Age of Irony Comes to an End." *Time* (September 24): 79.

Rosenfield, Israel. 1986. "Neural Darwinism: A New Approach to Memory and Perception." *New York Review of Books* 33 (October 9). www.nybooks.com/articles/1986/10/09/neural-darwinism-a-new-approach-to-memory-and-perc/

Ross, Andrew. 1991. *Strange Weather: Culture, Science, and Technology in the Age of Limits*. London: Verso.

Ross, Andrew, Ed. 1996. *Science Wars*. Durham, NC: Duke University Press.

Rouse, Joseph. 1993. "What Are Cultural Studies of Scientific Knowledge?" *Configurations: A Journal of Literature, Science and Technology* 1 (1): 1–22.

Rousseau, George S. 1978. "Literature and Science: The State of the Field." *Isis* 69 (4): 583–591.

——. 1987. "The Discourse(s) of Literature and Science." *University of Hartford Studies in Literature* 19: 1–24.

Rubery, Matthew. 2020. "The Confessions of a Synesthetic Reader." *Configurations: A Journal of Literature, Science, and Technology* 28 (3): 333–358.

Ruse, Michael. 1994. "Struggle for the Soul of Science." Review of *Higher Superstition*, by Paul R. Gross and Norman Levitt. *The Sciences* 34 (6): 39–44.

Saramago, José. 2004. *The Double*, translated by Margaret Jull Costa. Orlando, FL: Harcourt.

Sayers, Dorothy L. 1932. *Have His Carcase*. New York: Avon Books.

Sayers, Dorothy L. 1936. *Gaudy Night*. New York: Avon Books.

Sayers, Dorothy L., and Robert Eustace. 1930. *The Documents in the Case*. New York: Harper Paperbacks.

Scerri, Eric. 2007. *The Periodic Table: Its Story and Its Significance*. Oxford, UK: Oxford University Press.

Schachterle, Lance. 1990. "The Metaphorical Allure of Modern Physics: Introduction." In Slade and Lee 1990, 177–184.

Schneider, Eric D., and Dorion Sagan. 2005. *Into the Cool: Energy Flow, Thermodynamics, and Life*. Chicago: University of Chicago Press.

Shapin, Steven. 1980. "A View of Scientific Thought." *Science* 207 (March 7): 1065–1066.

———. 2019. "A Theorist of (Not Quite) Everything." Review of *Helmholtz: A Life in Science*, by David Cahan. *New York Review of Books* 66 (October 10): 29–31.

Shapin, Steven, and Simon Schaffer. 1985. *Leviathan and the Air-Pump: Hobbes, Boyle, and the Experimental Life*. Princeton, NJ: Princeton University Press.

Shelley, Mary. (1818) 2017. *Frankenstein: The Modern Prometheus*. In Guston *et al.* 2017, 1–187.

———. (1831) 2017. "Introduction to Frankenstein." In Guston *et al.* 2017, 189–193.

Shepherd-Barr, Kirsten. 2006. *Science on Stage: From Doctor Faustus to Copenhagen*. Princeton, NJ: Princeton University Press.

Shepherd-Barr, Kirsten, Ed. 2020. *The Cambridge Companion to Theatre and Science*. Cambridge, UK: Cambridge University Press.

Simon, Ed. 2018. "Is Frankenstein's Monster the Golem's Son?" Tabletmag. com, www.tabletmag.com/jewish-arts-and-culture/273792/is-frankensteins-monster-the-golems-son. Accessed December 9, 2019.

Sismondo, Sergio. 2017. "Post-Truth?" *Social Studies of Science* 47 (1): 3–6.

Slade, Joseph W. 1990. "Beyond the Two Cultures: Science, Technology, and Literature." In Slade and Lee 1990, 3–16.

Slade, Joseph W., and Judith Yaross Lee, Eds. 1990. *Beyond the Two Cultures*. Ames, IA: Iowa State University Press.

Slater, Matthew H., and Andrea Borghini. 2011. "Introduction: Lessons from the Scientific Butchery." In *Carving Nature at Its Joints: Natural Kinds in Metaphysics and Science*, edited by Joseph Peim Campbell, Michael O'Rourke, and Matthew H. Slater, 1–32. Cambridge, MA: The MIT Press.

Smith, Zadie. 2000. *White Teeth*. London: Penguin Books.

Snow, Charles P. 1956. "The Two Cultures." *New Statesman and Nation*, October 6: 413–414.

———. (1959) 1998. *The Two Cultures*. Cambridge, UK: Cambridge University Press.

———. 1963. *The Two Cultures: A Second Look*. Cambridge, UK: Cambridge University Press.

Soai, Kenso, Itaru Sato, Takanori Shibata, Soichiro Komiya, Masanobu Hayashi, Yohei Matsueda, Hikaru Imamura, Tadakatsu Hayase, Hiroshi Morioka, Hayami Tabira, Jun Yamamoto, and Yasumori Kowata. 2003. "Asymmetric Synthesis of Pyrimidyl Alcohol Without Adding Chiral Substances by the

Addition of Diisopropylzinc to Pyrimidine-5-Carbaldehyde in Conjunction with Asymmetric Autocatalysis." *Tetrahedron: Asymmetry* 14 (2): 185–188.

Soai, Kenso, Takanori Shibata, and Itaru Sato. 2000. "Enantioselective Automultiplication of Chiral Molecules by Asymmetric Autocatalysis." *Accounts of Chemical Research* 33 (6): 382–390.

Society for Literature, Science and the Arts. 2018. "Out of Mind: Program of the 32nd Annual Meeting of the Society for Literature, Science and the Arts." http://litsciarts.org/slsa18/wp-content/uploads/2018/11/SLSA-schedule-final.pdf. Accessed April 9, 2019.

Sokal, Alan D. 1996a. "A Physicist Experiments with Cultural Studies." *Lingua Franca* 6 (May/June): 62–64.

———. 1996b. "Transgressing the Boundaries: Towards a Transformative Hermeneutics of Quantum Gravity." *Social Text* 46/47 (Spring/Summer): 217–252.

———. 1998. "What the Social Text Affair Does and Does Not Prove." In Koertge 1998, 9–22.

Sokal, Alan, and Jean Bricmont. 1998. *Fashionable Nonsense: Postmodern Intellectuals' Abuse of Science*. New York: Picador USA.

St. Aubyn, Edward. 2012. *At Last: The Final Patrick Melrose Novel*. New York: Farrar, Straus and Giroux.

Stengers, Isabelle. 2018. "Science Fiction to Science Studies." In Meyer 2018a, 25–41.

Stokstad, Erik. 2019. "After 20 Years, Golden Rice Nears Approval." *Science* 366 (November 22): 934.

Stoppard, Tom. 1988. *Hapgood*. London: Faber & Faber.

———. 1993. *Arcadia*. London: Faber & Faber.

———. 1994. "Playing with Science." *Engineering & Science* 58 (1): 2–13.

———. 2015. *The Hard Problem*. New York: Grove Press.

Supran, Geoffrey, and Naomi Oreskes. 2017. "Assessing ExxonMobil's Climate Change Communications (1977–2014)." *Environmental Research Letters* 12 (8): 084019.

Swenson, Rod. 1992. "Autocatakinetics, Yes—Autopoiesis, No: Steps Towards a Unified Theory of Evolutionary Ordering." *International Journal of General Systems*, 21: 207–228.

Thomas, Lewis. 1983. "On Matters of Doubt." In *Late Night Thoughts on Listening to Mahler's Ninth Symphony*, 156–163. New York: Viking Press.

Tolstoy, Leo. 1869 (1968). *War and Peace*, translated by Ann Dunnigan. New York: New American Library.

Turkle, Sherry. 2009. *Simulation and its Discontents*. Cambridge, MA: MIT Press.

University of Bremen, n.d. "Fiction Meets Science." www.fictionmeetsscience.org/ccm/navigation. Accessed April 9, 2019.

University of Liverpool. n.d. "Literature and Science Hub." www.liverpool.ac.uk/literature-and-science/. Accessed April 9, 2019.

University of Notre Dame. n.d. "Faculty by Area." https://english.nd.edu/people/faculty/faculty-by-area/. Accessed April 19, 2021.

van den Broeck, Raymond. 1981. "The Limits of Translatability Exemplified by Metaphor Translation." *Poetics Today* 2 (4): 73–87.

van der Laan, Anna L., and Marianne Boenink. 2015. "Beyond Bench and Bedside: Disentangling the Concept of Translational Research." *Health Care Analysis* 23 (1): 32–49.

van Vogt, Alfred E. 1950. *The Voyage of the Space Beagle*. New York: Orb Books.

Wallace, David Foster. 1996. *Infinite Jest*. Boston: Little, Brown.

Watson, James D. 1968. *The Double Helix: A Personal Account of the Discovery of the Structure of DNA*. New York: Simon & Schuster.

Watson, James D., and Francis H. Crick. 1953. "Molecular Structure of Nucleic Acids: A Structure for Deoxyribose Nucleic Acid." *Nature* 171: 737–738.

Weinberg, Steven. 1994. "Response to Steve Fuller." *Social Studies of Science* 24, 748–750.

———. 2017. "The Trouble with Quantum Mechanics." *New York Review of Books* 64 (January 19). www.nybooks.com/articles/2017/01/19/trouble-with-quantum-mechanics/

Weininger, Stephen J. 1989. "Introduction." In *Literature and Science as Modes of Expression*, edited by Frederick Amrine, *xiii–xxv*. Dordrecht: Kluwer Academic Publishers.

———. 1990. "Concept and Context in Contemporary Chemistry." In Slade and Lee 1990, 39–49.

———. 2001. "SLS History." In *Conference Announcement of the Second European Conference of the International Society for Literature and Science*.

Wells, Herbert G. 1895. *The Time Machine*. New York: Henry Holt and Company.

White, Eric Charles. 1991. "Negentropy, Noise and Emancipatory Thought." In *Chaos and Order*, edited by N. Katherine Hayles, 263–277. Chicago: Chicago University Press.

Whitehead, Alfred North. 1925. *Science and the Modern World*. New York: Macmillan.

Whitworth, Michael H. 2001. *Einstein's Wake: Relativity, Metaphor, and Modernist Literature*. Oxford, UK: Oxford University Press.

Wigner, Eugene P., and R. A. Hodgkin. 1977. "Michael Polanyi, 12 March 1891–22 February 1976." *Biographical Memoirs of Fellows of the Royal Society* 23: 412–448.

Wikipedia. n.d. "English Translations of Dante's *Divine Comedy*." http://en.wikipedia.org/wiki/English_translations_of_Dante%27s_Divine_comedy. Accessed February 2, 2014.

Wilkes, James, and Sophie P. Scott. 2016. "Poetry and Neuroscience: An Interdisciplinary Conversation." *Configurations: A Journal of Literature, Science, and Technology* 24 (3): 331–350.

Williams, Zoe. 2003. "The Final Irony." *The Guardian*, June 28. www.guardian.co.uk/weekend/story/0,3605,985375,00.html. Accessed October 23, 2019.

Willis, Martin. 2015. *Literature and Science: A Reader's Guide to Essential Criticism*. London: Palgrave.

Willis, Martin. 2018. *The British Society for Literature and Science*. Review of *The Cambridge Companion to Literature and Science*. www.bsls.ac.uk/reviews/general-and-theory/steven-meyer-the-cambridge-companion-to-literature-and-science/. Accessed December 23, 2020.

Wilson, Edward O. 1998. *Consilience: The Unity of Knowledge*. New York: Vintage Books.

———. 2005. "Foreword from the Scientific Side." In Gottschall and Wilson 2005b, *vii–xi*.

Winterson, Jeanette. 2019. *Frankisstein: A Love Story*. New York: Grove Press.

Wolpert, Lewis. 1992. *The Unnatural Nature of Science: Why Science Does Not Make (Common) Sense*. London: Faber & Faber.

Woodcock, John. 1978. "Literature and Science Since Huxley." *Interdisciplinary Science Reviews* 3 (1): 31–45.

Woolf, Steven H. 2008. "The Meaning of Translational Research and Why It Matters." *Journal of the American Medical Association* 299 (2): 211.

Wyatt, Edward. 2005. "Literary Novelists Address 9/11. Finally." *New York Times* (March 7): E1.

Young, Emma. 2000. "Mutant Bunny." *New Scientist* (September 22). www.newscientist.com/article/dn16-mutant-bunny/

Zencey, Eric. 1990. "Entropy as Root Metaphor." In Slade and Lee 1990, 185–200.

Zimmerman, Seth. 2003. *The Inferno of Dante Alighieri: A Rhymed Translation*. http://infernodante.com/HELLI.html#top. Accessed April 16, 2014.

Zunshine, Lisa. 2015a. "Introduction to Cognitive Literary Studies." In Zunshine 2015c, 1–9.

———. 2015b. "Lying Bodies of the Enlightenment: Theory of Mind and Cultural Historicism." In Zunshine 2015c, 115–133.

Zunshine, Lisa, Ed. 2015c. *The Oxford Handbook of Cognitive Literary Studies*. New York: Oxford University Press.

Index

Printed in the United States
by Baker & Taylor Publisher Services